Cunningly Smart Phones

Cunningly Smart Phones

Deceit, Manipulation, and
Private Thoughts Revealed

Jack M. Wedam

Copyright © 2015 by Jack M. Wedam.

Library of Congress Control Number: 2015910181
ISBN: Hardcover   978-1-5035-8103-6
               Softcover   978-1-5035-8104-3
               eBook   978-1-5035-8105-0

All rights reserved. No part of this book may be reproduced or transmitted in any form or by any means, electronic or mechanical, including photocopying, recording, or by any information storage and retrieval system, without permission in writing from the copyright owner.

Cover Design by Nathan Hartley Maas

I want to thank Nathan for designing an excellent book cover. I also thank him for his patience as we worked through many styles and concepts to get a book cover that visually portrays the essence of this book.

Any people depicted in stock imagery provided by Thinkstock are models, and such images are being used for illustrative purposes only.
Certain stock imagery © Thinkstock.

Print information available on the last page.

Rev. date: 06/30/2015

To order additional copies of this book, contact:
Xlibris
1-888-795-4274
www.Xlibris.com
Orders@Xlibris.com

705940

Contents

Acknowledgements ix

Preface xi

Introduction xv

A Cunningly Smart Spy 1

Mary Godwin 8

What's Emotion Got to Do with It? 12

Reading Your Emotions and Targeting You 17

Playing with Emotions Is Very Profitable 27

Manipulating with Fear 32

The "Belongings Test": What You Buy,
    and Where You Go, Reveals More than You Realize 37

Big Data Gets to Know You,
    One Very Small Piece of Information at a Time 44

"Anonymous," but Very Well-Known 49

The Genius of Facebook 59

The Eyes Are the Window to the Mind 66

Electronic Mind Readers? 84

Can A "Mechanism" Accurately Reflect Your Thoughts? 86

"Wearables": For Your Health or for Their Profit? 95

Someday Soon, Maybe Very Soon 98
Legal and Ethical Issues 118
What Can You Do? 121
Glossary 125
Bibliography, References, and Websites 133
    Listed by Author's First Name 133
    Listed by Government Agency 161
        Department of Homeland Security 161
        United States District Court, Northern California,
            San Jose Division (litigation) 161
        Federal Trade Commission 161
        Office of the Under Secretary of Defense for
            Acquisition, Technology, and Logistics 162
        Securities and Exchange Commission 162
        US Attorney General 162
    Listed by Patent (assignee, inventor, or title,
        whichever is likely more relevant to the reader) 163
    Listed by Subject or Topic 165
    Listed by Website 168
Notes 177

This book is dedicated to my mom, who started badly but finished well. It is also dedicated to my dad, who guided me, loved me, fed me, nurtured me, and taught me well.

Acknowledgements

I want to acknowledge the many excellent reviewers whose comments made this book much more readable and understandable than my rough manuscript. This book would not have come to fruition without Audrey Charles. She read several rough drafts of my manuscript and provided many suggestions that greatly improved the manuscript. I also want to thank several other reviewers who provided me with helpful suggestions. These include Andy Aguirre, R. Clare Layton, Steve and Patty White, Jason Pike, John Short, Dr. Jim Gansberg, Dr. John and Cathy Maas, Dr. Albert Wedam, and Norene Wedam.

PREFACE

On September 21, 2007, the Department of Homeland Security awarded a contract to Battelle Memorial Institute to build a future attribute screening technology (FAST) demonstration laboratory.[1] Headlines such as "Behavioral biometrics to detect terrorists entering U.S."[2] and "Israel startup uses behavioral science to identify terrorists"[3] announced that the field of behavioral biometrics was in full stride.

Allegedly, behavioral biometric technologies are more accurate than a lie detector. Evidently, some nongovernment entities noticed the research and development of these new technologies. A few years after 2007, several corporations commercialized similar technologies—and the paradigm shifted drastically. Whereas in the first wave of development, the aim of using behavioral biometric technology was to help governments identify potential terrorists and to result in enhanced security measures to protect the public, the next wave of development, in stark contrast, saw the emergence of technologies that enabled corporations to *target* the public. The result of the latter was enhanced profits for corporations at your expense and an invasion of your privacy on an industrial level.

* * *

Allow me to discuss what qualifies me to write this book. I learned about neuroanatomy and neurology while studying to be of doctor of veterinary medicine more than three decades ago. I and my fellow students were required to know a great deal about neuroanatomy and neurology, which enabled us to conduct thorough neurological examinations of our patients—dogs, cats, horses, and other animals. The results of those examinations led to greater accuracy in making a diagnosis of disease or damage involving the nervous system. For example, abnormal eye movements and the eye's pupil response to light provide important information to help detect the location of damage to the brain.

Scientists have only recently revealed the function of saccades in the normally functioning brain. Saccades are very fast eye movements that allow the subconscious to monitor what is in the peripheral vision. When the function of saccades became known, I was fascinated. As I conducted research to discover more details, I learned how saccades and pupil responses could be used to probe the subconscious mind without the awareness or consent of the person being targeted. Intrigued, I decided to search further.

Mathematical sociology may sound like an awkward combination of disciplines, but it enables corporations to extract detailed information about you by analyzing your activity on social media websites. Mathematical sociology uses techniques that are similar to those used in epidemiology.[4] I earned a master's degree in veterinary preventive medicine, which included the study of epidemiology. My understanding of epidemiology enabled me to understand how mathematical sociology can be used to harvest valuable information about people from their social activities and participation with social media and social networks.

An understanding of patents and the patenting process has greatly facilitated my research into these new technologies. I was awarded a patent[5] for a finger-actuated electronic control apparatus. I am not a patent attorney. However, in the process of obtaining a patent I learned that patents provide a wealth of information about new technologies. Inventors must explain in excruciating detail what their patent is about and how their device or process is different from other devices or processes.

I discuss numerous patents in this book; however, you do not have to accept my explanations of them. I provide many references and links so you can read the patents yourself and make your own inferences. Alternatively, you can hire patent attorneys, neuroscientists, and other experts to explain the patents to you.

* * *

In this book, I reference material from two of my previous books, *Google Glass Can Read Your Mind* (2014) and *Billionaires and eGenies are $elling You Out* (2013). This notice serves as a general citation of my other books.

Introduction

I need your help.

We need to spread the word about consumers' Internet privacy. I hope this book will galvanize the public into action, the same way Ralph Nader's *Unsafe at Any Speed* inspired people to act. Because of Nader's book, a nationwide groundswell of people clashed with the titans of the automobile industry, which were then as powerful as the titans of the Internet are today. The public ultimately prevailed against the titans of the automobile industry. The result was safer automobiles and fewer highway fatalities. If enough people demand change this time, then we can prevail and take back control of our rights and privacy.

Together we can push back and force change. This is your chance to change things for the better. You can help yourself and others by pushing back and opposing the titans of the Internet who are trampling on our privacy. If you are a little hesitant about subscribing to the ideas presented in this book, then please make up your mind *after* reading it, not before.

I am a proponent of an ethical capitalist system in the United States. I believe that when corporations are deceptive, they deserve to be called out. I also want smartphone users to be aware of how smartphones and smartwatches may soon be able to be used in ways that few people would welcome.

I opted to keep the main body of this text to about half the length of other books that address Internet privacy, information technology, etc. As such, the main body of this text is fewer than thirty-five thousand words, not including the glossary, the bibliography, and the notes. A book of this length should appeal to readers who have an interest in their privacy but who don't have the time to read the longer books on the subject. Therefore, this book is not meant to be an exhaustive study of the many very technical fields of science and mathematics that have recently become interwoven with the pursuit of profits. Instead, this book is meant to be an overview to aid the reader in connecting the dots between mathematics (statistics), cryptanalytics, neuroscience, psychology, and marketing.

I struggled to simplify complex topics in order to make this book readable at the twelfth-grade level.[1] In order to achieve that objective, I avoided discussing many of the complex details of various topics in the main text. Some reviewers asked me to include more details, whereas other reviewers asked me to include fewer details. To satisfy both groups, I moved some explanations from the main body of the text to the notes so that the information would be available to those who wanted more detail. A glossary is included to provide more information. This book also has a long bibliography that contains many references. You may access these references if you want to learn more.

Almost half of this book is devoted to the glossary, the bibliography, references, and notes. This is a large amount of space devoted to these areas. As mentioned above, it moves much of the complex discussion out of the main body of the manuscript and allows the reading to flow smoothly for the rapid reader who does not want to bother with the details. More importantly, *Cunningly Smart Phones* provides an abundance of evidence to support assertions that may be denied by the titans of the Internet. The titans of the Internet are very good at making accurate inferences about you. This book will provide you with ample information so you can make accurate inferences about their activities and also come to see how their activities might affect you in ways you may not welcome.

A Cunningly Smart Spy

Your smartphone is *cunningly* smart and is quickly becoming even more cunning. Your smartphone is spying on you and sharing many of your secrets.[1] Your smartphone is eavesdropping on you, using very crafty methods to do so. We typically think of spying as activity directed against our enemies. However, things have drastically changed in recent years. Espionage-related tools and techniques that were developed to defend us from our enemies are now being used by many corporations to spy on us. Some of these have been developed by the Defense Advanced Research Projects Agency (DARPA) or its forerunner, the Advanced Research Projects Agency (ARPA).

Smartphone and other technology companies are gathering huge amounts of your personal information and thoroughly analyzing it. If you are wondering what difference this makes, then you will be interested in the chapter titled "Someday Soon, Maybe Very Soon," which illustrates how your personal information could soon be used to manipulate you. If you do not like the notional scenarios of manipulation, then you may read about several real cases of manipulation and economic discrimination in chapter 8 of Bruce Schneier's 2015 book *Data and Goliath*.[2]

Smartphones have various sensors that provide some clever ways to capture people's thoughts and emotions in ways that were unimaginable just a few years ago. After your personal information is gathered, it is quickly auctioned.[3] Electronic devices can now "accurately reflect a subject's actual thoughts."[4] Will your private thoughts soon be auctioned off without your awareness or consent?

Some Internet-based corporations know your likes and dislikes, your emotions, and even subtle changes in your mood.[5] They can pinpoint where you will be tomorrow[6] and even in future years.[7] Some corporations know very private details about you that you may not want them (or other people) to know.[8] These corporations "can more or less know what you're thinking about."[9] Marketers can use your information to target you with sophisticated advertisements. Marketing psychologists have learned how to manipulate consumers' brains to induce a state much like being in love. Having also learned how cocaine stimulates the brain, some marketing psychologists have learned how to stimulate the relevant part of the brain with advertisements.[10] This means that cravings can be implanted and stimulated in consumers.

Cocaine is illegal. Nevertheless, the sale of cocaine generates huge revenues. In stark contrast, new business models that bring Internet titans and marketing psychologists together are legal and are much more profitable than cocaine. The old adage "Follow the money" will help you make sense of many recent developments in Internet approaches. The old saying "There is no free lunch" needs to be updated. Insiders in Silicon Valley now say, "If the service is free, then you're the product."[11]

Many Internet-based corporations are scrambling[12] to find out more about you. In the process, they are trampling on your privacy much more than the US government is able to do. Even if government agencies ignored all the privacy laws and spied on you, they could not match the ability of Internet-based corporations to invade your privacy. It costs money to spy on you, store information about you and analyze your activities. Even spies are constrained by economic fundamentals, budgets, and economies of scale. The government cannot sell your information and, in so doing, recover the expense of their spying. In contrast, many corporations are making huge profits by spying on you and selling your secrets.

A *Wall Street Journal* investigation found that "one of the fastest-growing businesses on the Internet is the business of spying on consumers." The *Wall Street Journal* published a series of articles beginning with "The Web's New Gold Mine: Your Secrets,"[13] which included many fascinating and disturbing details. The Federal Trade Commission has concluded that Google has done "real harm to consumers."[14] Can we trust corporations like Google? The technologies that spy on you did not just suddenly and serendipitously sprout up in the past few years. Following is a brief historical review to provide you with some context and make it easier for you to understand the current situation.

Theodore Paraskevakos was an inventor who was awarded more than fifty patents worldwide.[15] He is generally credited with inventing the first modern smartphone by combining the telephone with some computational ability.[16] However, there is much more to the story. Computers in smartphones evolved from a long line of mechanical computers, the first of which was developed just as World War II began. Alan Turing developed the bombe[17] for Great Britain. The bombe, a cryptanalytic machine that deciphered secret messages encoded by the Enigma device in Germany during the era of World War II, was featured in the 2014 movie *The Imitation Games*. This bombe was a real game changer, as it ushered in electronic computers. Now, seven decades later, smartphones have risen to the apex[18] of a nexus consisting of scientific disciplines such as mathematics, sociology, psychology, neuroscience, and artificial intelligence. These are all integrated into programs run by computers that are much more sophisticated than those capable of deciphering Enigma messages.

The Economist ran an article titled "Smartphone security: The spy in your pocket," which revealed how spies working in the Government Communications Headquarters (GCHQ), Great Britain's equivalent of the National Security Agency (NSA), had stolen encryption keys to SIM (subscriber identity module) cards, which allowed spies to quickly decrypt conversations and data from smartphones.[19] Although this news may shock some people, it should not be shocking, because corporations are already gathering much more data from smartphones than are government spies. With the new iOS9, Apple's new iPhone will be able to tell exactly how often, when and how iPhone users have sex.[20] Smartwatches can determine when a person is sleeping or having sex.[21] (Can you trust corporations with that information?) Moreover, corporations soon will have tools so sophisticated that they will be able to gather even more information from more people and to make more accurate inferences about you than government spies ever could. That is unless corporations share their technology with the governments spies or the government spies hide under your bed and following you around all day with teams of interrogators and psychologists.)

Technology now exists that can quickly find out details about people that are so private and sensitive that many people would be uncomfortable sharing them with others. Some people are afraid of spiders, snakes, bats, etc., but are ashamed to admit it. Others are afraid of appearing to be less than competent. Some people are afraid of lacking control in some areas of their lives. Having insight into your fears and apprehensions provides marketers with tremendous advantages, as they then have the ability to sell you something to help you feel better.

Most individuals are aware that many corporations scoop up vast amounts of information about people. Many are deeply suspicious of what corporations are doing with their personal information.[22] According to the Pew Research

Center, the number of adults who agree that consumers have "lost control over how personal information is collected and used by companies" is larger than the number of adults who are concerned about the "government accessing some of the information they share on social networking sites without their knowledge."[23] Yet, some have become indifferent to calls urging for using more caution and being provided with more protection. Others have become indifferent because the warnings are as "inevitable in the digital age as death and taxes"[24]—so why worry? Cautions about Internet privacy can "seem more tiresome than threatening."[25] Some people rebel against cautions in the same way that children rebel against parents who nag them and say, "Go clean your room!"

Other people have concerns about privacy but are willing to make some trade-offs in exchange for "free" services.[26] Some push privacy concerns aside because they believe that they have nothing to hide. Furthermore, many of those who have been recently enriched by gathering, selling, or analyzing private information accuse those who are concerned about Internet privacy of being crazy, uninformed, ignorant, and paranoid—conspiracy theorists, right-wing nuts, left-wing nuts—or various combinations.

By reading the facts presented by various experts, patents, scientific papers, official government documents, and court proceedings, you can form your own opinion. Your having read this far indicates that you have the ability to use critical thinking to evaluate the complex issues presented by monitoring technology. Good for you. Pat yourself on the back for having an open mind concerning your smartphone use.

* * *

A 2014 Federal Trade Commission report states, "In today's economy, Big Data is big business."[27] Many credible sources indicate that your personal information (and the analysis of your information) is like a gold mine.[28] "The Internet is the most fantastic source of data the world has ever seen," one that creates huge potential for earning profits.[29] However, are the profits legitimate?

On April 15, 2015, the European Union accused Google of "cheating consumers and competitors."[30] That should not surprise anyone, since Google CEO Eric Schmidt "agreed that Google has the kind of dominant market share that might subject it to monopoly rules."[31]

On the other hand, some people have used the Internet to ensure Internet users' privacy. Mr. Michael Fertik founded Reputation.com in order to stop cyberbullying.[32] According to Mr. Fertik, the current business model for Internet-based companies involves corporations' giving customers something for free and then gathering information on them and selling that information to third parties. Mr. Fertik proposes that the business model should be changed so that people have the opportunity to decide if they want to sell information about themselves to corporations.[33] He proposed a good idea, but it did not get far. Mr. Fertik's proposal would have required that many multibillion-dollar corporations, millionaires, and billionaires share their newly acquired wealth with you.

Many people do not know how much information is being scooped from them by way of deceptive methods. Furthermore, most people do not know how much their information is worth. Analyzing the revenues from just five corporations in 2013 indicates that personal information was worth at least $43 per person in the United States. That figure was probably around $50 for 2014.[34] (Including more than just five corporations would increase that number.) That is not chump change unless you are a person who has bundles of money to throw away.

What would you do if you were able to sell your private information and your private thoughts? Would you sell some or all? What would you do with an extra $50 a year paid to you in exchange for your private information and thoughts? Would you save it? Would you buy something for yourself? Would you buy a gift for someone?

Before smartphones were invented, the main ways to reach people were by using TV, radio, and print media. However, these methods were sometimes ineffective. John Wanamaker, a US department store merchant, lamented, "Half the money I spend on advertising is wasted; the trouble is I don't know which half."[35] Modern technology now allows smartphones to learn much about you and to provide instantaneous feedback about the effectiveness of advertisements. Advertising has gone beyond showing consumers a variety of goods and services they may be interested in buying. Advertisements enabled by smartphones can now manipulate people's emotions, leading them to waste their money on certain goods and services. The recently developed crafty advertising methods are so effective that CEOs of some companies that use the techniques are very concerned about having their brands associated with brain manipulation.[36]

Before we jump into the details, let's take a quick look at history in order to help put things into perspective. In the next chapter, we see that concepts that were once in the realm of science fiction have been turned into huge moneymakers.

Mary Godwin

On May 14, 1816, an eighteen-year-old Englishwoman, Mary Godwin, along with her family and friends, arrived in Geneva, Switzerland, for a vacation. However, 1816 was a year without a summer, as the world was enduring a volcanic winter after the eruption of Mount Tambora in 1815. The cold weather and incessant rain kept Mary Godwin and her party indoors for days. They passed their time telling ghost stories. Mary did not have any ghost stories to tell. Every morning after the first person told a tale, Mary was asked if she had a ghost story to share. Sad and chagrined that she had nothing to share, she set out to write what she thought would be merely a short essay so she could avoid the uneasiness she felt when she reported that she had nothing exciting to add to the group's conversation.

However, once Ms. Godwin caught hold of an idea, she created much more than just a short essay. She wrote a novel. Two years later (1818), at the age of twenty, she anonymously published her novel in England. Many consider her novel to be the first and the most successful science fiction novel of all time. Scores of movies and other books reflecting Mary's original theme have been, respectively, made and written. A few lines in Mary Godwin's novel have escaped the realm of science fiction and have become reality. What was amazing but impossible for almost two centuries now looks prophetic in hindsight. Many technologies that were newly developed at the time and that appear in Godwin's novel seem bland compared to the technologies that can now accomplish things on an industrial level.

In Mary's novel, the character named Victor creates a creature. In due time, Victor and his creature converse. Victor's creature offers a few comments. Do you recall what he said? This is part of the dialogue, in case you have never read the original text or have forgotten: "What chiefly struck me was the gentle manners of these people ... I would remain quietly in my hovel, watching, and endeavouring to discover the motives which influenced their actions."[1]

Later, the creature says, "My thoughts now became more active, and I longed to discover the motives and feelings of these lovely creatures."[2]

Godwin is Mary's maiden name. It was her surname when she wrote the first draft of her now famous novel. Perhaps you recognize her better by her married name—Mary Wollstonecraft Shelley. Victor's creature is more commonly known as Frankenstein.

Victor made the creature by stitching together organic pieces and parts of living things. We now have electronically created creatures, ones that include several technologies that are electronically stitched together and interwoven so seamlessly that they can operate together autonomously. These technologies include communications devices,[3] hardware such as computers and servers, data storage devices,[4] software,[5] and artificial intelligence. Do not worry if the concept of electronically created creatures seems a bit vague now. We will come back to it later in the book and achieve clarity.

Science and technology have advanced in ways that Mary Shelley could not have imagined. The information that follows in this book will show how science and technology have enabled some very clever people to create electronic entities that can behave like Frankenstein and discover the motives that influence people's actions. In some cases, these electronic creations can learn more about people than they know about themselves. Technology now exists that allows computers and autonomous algorithms to probe the subconscious mind and ferret out things of which people are not consciously aware. Alternatively, if people were aware of such thoughts, then they may be hesitant to admit those thoughts to family and friends—or even to themselves.

Smartphones now serve as links between a person's brain and these electronically created entities. Consider the contrast. Frankenstein was a bumbling oaf who was trying to understand and befriend people, yet he was exquisitely effective at frightening people. In comparison, modern smartphones, apps, algorithms, and other electronic devices and procedures are *superexquisitely* effective at discovering people's "motives and feelings."[6]

Surprisingly, smartphones elicit little of people's concern. Perhaps an explanation for this is the comparison of that which can be seen and that which cannot be seen. The villagers in *Frankenstein* were terrified by the creature's monstrous appearance, but with the smartphone there is nothing to see beyond the device itself. All of the information gathering, information analysis, and individual targeting goes on quietly behind the scenes and is completely invisible to the person holding the smartphone. Furthermore, promoters actively tamp down people's fears and concerns. To alleviate fears, promoters use phrases like "enhanced Internet experience" or "more enjoyable Internet experience."

How do marketers play with people's emotions? The first step is to find out what the targeted individual is thinking about at both the conscious and the subconscious level. Marketers then find out how to elicit an emotional response in the targeted individual. This is where modern science has eclipsed *Frankenstein*. Big Data is using smartphones to do what Frankenstein was unable to achieve. Although Frankenstein endeavored "to discover the motives which influenced their actions"[7] and "longed to discover the motives and feelings of these lovely creatures,"[8] he failed miserably. He was only able to frighten people. Whereas Frankenstein could only scare people away, the modern science of fear has shown marketers how to frighten consumers and lead them to buy products and services that they may not really want or need. Modern science has shown marketers not only how to manipulate with fear but also how to implant and stimulate positive emotions in people so that they will buy the products and services on offer.

Perhaps you think this analogy to *Frankenstein* is over the top. Paul Ekman, PhD, is an eighty-year-old psychologist who added a tremendous amount to the work of others who studied emotions and their influence on facial expressions.[9] He is credited with developing the facial action coding system (FACS),[10] which is the basis of the TV drama *Lie to Me*. In a *Wall Street Journal* article, "The Technology that Unmasks Your Hidden Emotions," the writers say that Dr. Ekman "fears he has created a monster."[11]

Possible monsters aside for now, why is there so much corporate interest in unmasking people's emotions?

What's Emotion Got to Do with It?

The answer to the question that serves as this chapter's title comes from a leading neuroscientist, Joseph LeDoux, PhD. This eclectic man has many talents. In addition to being a gifted neuroscientist who has written several books,[1] he is a songwriter, a singer, and a guitarist in a rock band, the Amygdaloids, that has produced several CDs. Moreover, he has performed at the Kennedy Center in Washington, DC, and in a concert at Madison Square Garden, playing at the latter venue to a crowd estimated at ten thousand.[2] Dr. LeDoux was one of the first people to understand emotions from a scientific perspective. In 1996, he published *The Emotional Brain: The Mysterious Underpinnings of Emotional Life,* which was groundbreaking for its time and which has served to establish a firm scientific foundation for understanding how emotions influence behavior.

A study of effective ads found that "campaigns with primary emotional content performed twice as well as the approaches that focused on rational content."[3] The question of whether the emotion is a result of a person's core values or is planted in the brain as the result of well-designed advertisements is almost irrelevant. Once people's emotions are stimulated, or once emotions are implanted by advertisements, many people will act out those emotions. The words *emotion* and *motivation* are both derived from the same Latin root, *movēre*, which means, "to move."[4] This implies an association that links emotion and action.[5] Indeed, Dr. LeDoux found that "once emotions occur they become very powerful motivators of future behavior."[6] Perception Research Services International, Inc., makes this comment about measuring emotions in consumers: "Measuring their emotions is especially useful in those cases in which they may be unwilling or unable to articulate their true feelings."[7]

However, the ability to peek into the mind is just the first step in sophisticated marketing (or manipulation). According to George Lowenstein of Carnegie Mellon University, as quoted by Martin Lindstrom in *Buyology*, "Most of the brain is dominated by automatic processes, rather than deliberate thinking. A lot of what happens in the brain is emotional, not cognitive."[8] Alternatively, one company states it more succinctly: "Emotions drive action."[9] Furthermore, by measuring potential consumers' brain responses to various emotional appeals made in advertisements, marketers can continually refine the emotional appeal until it elicits a strong emotional response in their targeted demographics. Research recently showed that decisions are often contingent on emotions[10] and that "emotions exist to guide behaviour."[11] Therefore, if a strong emotion can be elicited in a person, then that person is likely to take action based on his or her emotions.

Emotions play a large role in persuading us, whether we are willing to admit it or not. More important, though, is how marketers get to know us and map our emotions. Once you understand that process, you will have a better understanding of why personal information is very valuable to marketers. Playing with people's emotions has become big business because it delivers big profits.

* * *

Scientists have accurately determined which parts of the brain are associated with various mental activities, such as feeling emotions, hearing, seeing, lying, and telling the truth. One electronic device can determine within a few seconds any increases in a person's brain activity. This device monitors how much oxygen is being used in different parts of the brain.

Scientists and medical practitioners prefer the acronym—fMRI[12]—instead of the long name of this electronic device (functional magnetic resonance imaging). Unfortunately, neither the acronym nor the long name lends itself well to our understanding of why this device is useful to studying the brain's activity. If the brain cells are more active than usual, then those brain cells need more oxygen, because brain cells utilize oxygen just like muscles and every other cell in the body do. This device can detect areas of the brain that are using more oxygen than they would if they were not stimulated. It can "see" areas of increased activity at a resolution slightly smaller than one-tenth of a millimeter[13] (about four one-thousandths of an inch). The resolution is five times narrower than the 0.5 mm lead used in some mechanical pencils. Therefore, this device can pinpoint small, specific parts of the brain that are active.

The fMRI shows which part of the brain is processing input coming from the body's various sensors, such as the eyes, ears, nose, taste buds, skin, etc. More than merely "seeing' that the brain is "thinking" about something, fMRI can provide valuable insight into *what* someone is thinking. Since neuroscientists have linked specific functions of the brain with specific areas of the brain, the fMRI can accurately tell if a person is thinking about something logically or emotionally. If a person is telling the truth, then a certain part of his or her brain lights up on the fMRI display screen. On the other hand, if someone is trying to lie, then a different spot of his or her brain lights up on the screen. According to a 2007 story in *The New Yorker,* the Defense Advanced Research Projects Agency (DARPA) funded much of the research of the fMRI for use as a lie detector test by the Department of Defense Polygraph Institute.[14] Two companies offer lie detector tests based on this technology.[15]

More than just telling if someone is lying or telling the truth, fMRI can provide insight into how someone interprets images and messages. It appears that many corporations noticed that DARPA's research and development was paid for by taxpayers. In his *New York Times*–listed best seller *Buyology* (2010), Martin Lindstrom describes the fMRI as "a killer marketing tool."[16] Lindstrom also offers many fascinating examples of how marketers use the fMRI to refine their advertising campaigns. Writing for *Bloomberg,* Amber Haq summed it up well: "Neuromarketers use sophisticated brain-imaging technology to test consumer response and help clients fine-tune their strategies."[17]

* * *

Benjamin Libet of the University of California, San Francisco, found that there was a time lag between the time when the mind was *ready to make a decision* and when the mind *actually made that decision* and the person took action.[18] Using different technology, researchers in Leipzig, Germany, were often able to predict that a person had made a decision several seconds before the person was aware that he or she had made that decision.[19]

Marketing psychologists are trying to understand what takes place during the few seconds before consumers make a final decision. If they can intervene in the decision-making process before the consumer makes a final decision, then they will be more successful in manipulating consumers.[20] Defenders of these techniques will not call the practice manipulation. Instead, they carefully parse their words in statements similar to this one: "If we present that stuff in an engaging and fun way then people have a better experience and sometimes they buy more."[21]

You can decide for yourself whether these "interventions" constitute manipulation. You can also determine how these new technologies may affect you. You do not need "experts" whose paychecks depend on the profits from these new technologies to dismiss your legitimate concerns.

* * *

Some people appear to have fallen in love with their smartphones in almost the same way as people fall in love with other people. This ought not to be surprising. After all, smartphones are a readily available conduit that some people use to connect with those whom they cherish when having a face-to-face conversation is not practicable. Therefore, it is easy to see how the animate and inanimate can become intertwined.

Love is blind. As it is with personal relationships, so it may be with smartphones. Some are willing to overlook the problems that smartphones pose because they are blinded by their love of the device. As people can be betrayed by friends and lovers, people may find that their trust and feelings are betrayed by their smartphones.

Reading Your Emotions and Targeting You

The field of emotions analytics[1] is a hot new area of research and commercialization. Algorithms can reveal emotions through voice analysis.[2] Beyond Verbal,[3] a start-up company, claims to have designed smartphone algorithms that can infer[4] a person's emotional state by using a brief sample of the person's voice captured by his or her smartphone's microphone. The company claims to have the ability to analyze "emotions from vocal intonations" instead of by analyzing statements like "I'm happy" or "I'm sad." The algorithms are designed to detect subtle changes in the voice, such as pitch, loudness, pauses, etc., that are many of the same clues one uses to detect emotions in the voices of one's friends and family. The algorithms then compare the voice qualities to voice patterns in the company's database.[5] What you do intuitively, computers can now do autonomously. A computer listens closely to what you say and how you say it. Several companies are getting on board with this new emotions-analytics technology. Ginger.io uses "sensor data collected through the phone" to help detect if a person is depressed and needs help.[6]

Facebook recently bought the start-up company Wit.ai, which claims to "turn speech into actionable data."[7] However, Facebook's executives will not reveal exactly what actionable data they expect to obtain. Why is Facebook being secretive? In the absence of a clear indication of what "actionable data" it may glean from voices, we can reasonably assume that Facebook is ready to commercialize voice-based emotion analytics.

Smart TVs have microphones that can eavesdrop on you and on whomever else is in range of the microphone. After that, the conversation is data-mined.[8] Samsung's corporate policy indicates that the company listens in on the conversations of users of Samsung TVs: "Please be aware that if your spoken words include personal or other sensitive information, that information will be among the data captured and transmitted to a third party through your use of Voice Recognition."[9] Samsung's computers decode what is said by using voice recognition software.[10] Following Samsung's lead, Apple intends to get into the Web TV service.[11] This will provide Apple with yet more information to make inferences about you.

Bruce Schneier has written more than a dozen books dealing with computer and Internet security. In chapter 6 of his 2015 book *Data and Goliath,* he explains the "public-private surveillance partnership."[12] Similarly, President Eisenhower cautioned the American public about the military–industrial complex.[13] In this relationship industries earn profits from government purchases. In other cases, the government pays for the research of new technologies which corporations then commercialize. In some cases, the government pays for much of the research while corporations patent the commercial use of those technologies. Therefore, some corporations encourage the government to fund research on new technologies which the corporations know they will be able to earn profits. Have we forgotten Eisenhower's warning now that the funding for technology has been shifted from the space race to the moon and from fighting the former Soviet Union to identifying potential terrorists and fighting asymmetric warfare?

Apple acquired Siri,[14] a natural language app that is a spin-off technology developed under a five-year, $150 million

Defense Advanced Research Projects Agency (DARPA) effort.[15] Consider the fact that taxpayers paid $150 million to develop a technology that could be used to spy on them. Is this an example of what President Eisenhower warned us to be watchful? Apple has a well-orchestrated public relations campaign, which it leverages to make it seem like they are doing their best to protect you from government surveillance. In reality, Apple benefits from a taxpayer-funded project that has the potential to spy on the public.

USA Today ran an article explaining that the Drug Enforcement Administration (DEA) has, for many years, been tracking the telephone numbers of billions of calls. Although this may anger many people, it is important to note that the DEA does not record the content of those calls.[16] Apple recently filed a patent application for a voice-controlled intelligent digital assistant. Will Apple use this technology together with another technology (discussed below) that it has recently developed for the purpose of inferring users' moods? Whereas some people have become angry at the US government because it collects telephone numbers (but does not record the calls), not as many are angry at Apple, which has been quietly developing technology that can infer people's moods in addition to eavesdropping on them .

* * *

Lisa A. Williams, a lecturer at the School of Psychology of the University of New South Wales, Australia, wrote an article in March 2014 titled "Your smartphone is looking at you—but can it read your emotions?"[17] There is no question that large computers can read emotions. Smartphones can also read facial expressions in controlled conditions (i.e., in laboratories). The question is, when will your smartphone be reading your facial expressions to determine your emotions? The other important question is whether you will be supplied with a user agreement that you are able to understand. Will you sign up for a "free" service by clicking "Yes" (i.e., giving your consent) after reading a long and difficult-to-understand user agreement?

An early system of reading facial expressions was developed by Carl-Herman Hjortsjö, a Swedish anatomist. The more advanced system—the facial action coding system (FACS)—was developed[18] by the same Dr. Paul Ekman who was mentioned earlier. Ekman became interested in facial expressions in the 1960s, but he realized that there were many gaps in his understanding of them.

FACS likely would not have come to fruition had Dr. Ekman not had "two strokes of luck." He said, "Through serendipity the Advanced Research Projects Agency (ARPA) of the Department of Defense gave me a grant to do cross cultural studies of nonverbal behavior." The other stroke of luck was his meeting Silvan Solomon Tomkins, PhD,[19] a psychologist who was on staff at the Harvard Psychological Clinic and who taught psychology at Princeton and Rutgers.[20] Tompkins is described as brilliant by many, as is evident in the following description. He "was the author of *Affect, Imagery, Consciousness,* a four-volume work so dense that its readers were evenly divided between those who understood it and thought it was brilliant and those who did not understand it and thought it was brilliant."[21]

Tompkins came across his ability to read facial expressions accurately in a very unusual way. To earn money for his doctoral studies at Harvard, he worked for a syndicate that determined the handicap of horses (i.e., determined which horses were most likely to win or lose a race). He did this by sitting in the grandstands for hours and using binoculars to watch horses' faces and their interactions with other horses. Apparently, he was very good at this task, was paid very well for his efforts, and lived lavishly.[22]

This story resonates with me. As a veterinarian who worked with horses earlier in my career, I also learned to read horses' emotions from their facial expressions. I was twice able to escape serious bodily injury or death by being keen to a horse's subtle facial expressions and by knowing the difference between breeds, as some breeds have very short fuses. On rare occasions, horses will use a front hoof to strike forward with speed, power, and accuracy that far exceeds the ability of any professional boxer. The strike is lightning fast. If a horse's intent is malicious, then the target will not see the hoof coming. To my good fortune, twice I recognized the subtle early signs of explosive danger in two horses' facial expressions, a fraction of a second before the horses exploded. I felt the whoosh of air as a horse's hoof barely missed my ear on one occasion and barely missed my nose on another occasion. Silvan Tomkins, the doctoral student and extraordinary horse handicapper, may have lived lavishly, thanks to what he earned by having horse sense. I, however, am delighted that I did not suffer serious bodily injury or death after those two ill-tempered horses tried to strike me.

When Dr. Tomkins's son was born, he took a sabbatical to care for the boy. During that time, he spent many hours observing his son and was able to accurately identify many emotions, especially the very subtle prefacial expressions that last for only a fraction of a second before the more recognizable facial expression becomes apparent. As his son grew older, Tomkins added more emotions to his list. Building on this, he took thousands of pictures of human faces and catalogued them according to the emotion they displayed.[23] Dr. Ekman studied Dr. Tompkins's work and further developed it.

Recently, software has been developed that allows computers to use FACS to determine emotions from facial expressions. A new Japanese humanoid robot can determine people's emotions by analyzing their facial expressions.[24] Start-up companies are trying to develop apps to allow smartphones with smaller cameras to do what larger computers and larger cameras are doing currently. An emotion measurement technology company, Affectiva,[25] recently teamed up with a video-chat service, ooVoo LLC,[26] the latter of which serves a hundred million smartphone users, in order to design an app that can reveal emotions during video chats on smartphones.[27]

* * *

Apple Inc., famous for its iPhones, iPads, and iPods, recently filed a patent application for a technology that will allow Apple to infer your emotions while you are using its products.[28] Although the patent has not been issued yet, the patent-pending status allows Apple to use this technology without encountering competition from others. Therefore, Apple may be legally using this new technology to infer your emotions right now.

According to the recent patent application, Apple's system first establishes a baseline mood for the user and then applies mood rules to establish a mood profile for that user (page 12, claims 1–6). After this, the system utilizes the individual user's inferred mood profile to better target advertising for that person (pages 13–14, claims 7–27).

In this patent application, Apple's language is very coy. It carefully dodges offering explanations of how Apple gathers the "mood-associated characteristic data" about you. Instead, Apple simply offers it as a foregone conclusion that the company may gather as much mood-associated characteristic data as it wants. With all of your mood data, Apple can establish a solid baseline on you as well as detect subtle differences in your baseline.

It appears that Apple is planning to gather mood-associated characteristic data on users of Apple products. Perhaps Apple has been gathering mood data on you already. Many people were furious at the US government when Edward Snowden revealed that the government was gathering information such as the telephone numbers of calls made by citizens. Now it appears that Apple has been flying under the radar (or soon will be) and gathering mood-associated characteristic data on people. Here is a short excerpt from the patent application:

> A method of inferring mood can be based on deviations from a baseline mood profile. The baseline mood profile can be a general baseline mood profile representing a standard mood for a hypothetical person. However, such a baseline mood profile may not accurately represent each user. Therefore, individual baseline mood profiles can be generated for each user based on analysis of mood-associated characteristic data specific to the user. As more information is known about a user, a baseline mood profile can become more accurate, thus resulting in more accurate mood inferences.

If you want to know more details, then you should read more about this patent application at http://pimg-faiw.uspto.gov/fdd/20/2014/56/002/0.pdf. In this patent application, some form of the word *target* is used sixty-two times. Whom do you think will be among those targeted by Apple? How do you think people will respond when they realize that Apple considers them not to be users of its products but targeted individuals whose digital identity, mood data and other inferences can be sold to marketers?

Apple, as are other, similar companies, is very secretive about how much data it can store. Apple has several large data centers and is expanding its data storage capacity.[29] On February 2, 2015, the technology giant announced that it was spending approximately $2 billion to convert a former GT Advanced Technologies facility in Mesa, Arizona, into a data center.[30] Apple is secretive about its data storage capacity. "An

Apple spokeswoman did not specify how much data center space the facility will hold or what its power capacity will be."[31]

Data storage and data processing consume electrical power. Therefore, knowing the facility's power consumption would provide one with an effective method to estimate the data storage capacity. However, Apple is also withholding information about its power consumption. What is Apple trying to hide? Why will Apple not tell us how much data storage capacity it has at its disposal?

Seeing that Apple will not reveal the amount of data it can store, we can use an indirect method to estimate its data storage capacity based on money spent in certain areas. Although the mission of the Utah Data Center[32] is classified, some suspect that it holds data that the National Security Agency (NSA) was gathering on citizens. Apple's expenditure of $2 billion is one-third more than the amount spent by the NSA ($1.5 billion[33]) to build a data center in Camp Williams, Utah. It is a reasonable assumption that Apple gets a better deal than the US government since Apple buys more data storage hardware (Apple has several large data storage sites, whereas the US government likely has only one large data storage facility). Therefore, we can assume that Apple's new data center has at least one-third, and maybe more, of the data storage capability of the new NSA data center, which is estimated to store about ten exabytes.[34] How much is an exabyte? Verlyn Klinkenborg of the *New York Times* provides this comparison: "Five exabytes, as it happens, is equivalent to all words ever spoken by humans since the dawn of time."[35]

On February 23, 2015, Apple announced its plan to build two new data centers in Europe, which will cost approximately $1.9 billion.[36] Since these two new sites will cost more than the Utah Data Center, we can assume that they will also have more data storage capacity than the Utah Data Center.[37] Indeed, Apple claims that the new data centers in Europe "would be among the largest in the world,"³⁸ which provides another indication that the facility will be larger than the Utah Data Center.

Apple has far more storage capacity than is needed to store the offerings of iTunes. Remember Verlyn Klinkenborg's comparison: just five exabytes is enough to store "all words ever spoken by humans since the dawn of time." Therefore, we can assume that much of what Apple is storing is people's personal information. Moreover, Apple admitted that it is storing a lot of personal data when it stated that it wanted to build data storage centers in Europe to try to keep the NSA from gaining access to people's personal information.³⁹ It appears that Apple has a huge appetite for your personal information. This leads us to ask the question of what Apple is going to do with all of this personal information. Apple's patent on mood data may provide some insight into what Apple may do with your information.

It appears that Apple may soon have at least eight data centers.⁴⁰ As we have already seen, each of Apple's data centers is more likely to hold at least 30 percent more data than the Utah Data Center. In the absence of reliable information from Apple, we must make assumptions based on money spent. The eight Apple data centers will have ten times (or more) the data storage capacity of the NSA's Utah Data Center.

NSA will not say what data is being stored, and Apple is coy about what information it is storing. We know that Apple has submitted a patent application for a method of collecting and analyzing information about your mood. Google also has a huge amount of data storage, thirteen large data centers around the world, including six large data centers in the United States.⁴¹ To give you an indirect estimate of the data storage capacity by analyzing expenditures, Google has invested $700 million in its Mayes County, Oklahoma, data center,⁴² has invested $1.2 billion in its Berkeley County, South Carolina, data center,⁴³ and will have invested $2.5 billion by 2019 in its Council Bluffs, Iowa, data center.⁴⁴ Facebook leases data storage space from six different data centers in Silicon Valley, one facility in San Francisco, and three facilities in Ashburn, Virginia. Facebook has built a data center in Prineville, Oregon, and had plans to build data centers in Forest City, North Carolina, and Lulea, Sweden,[45] which should have been completed by the publication of this book.

Considering how much information these companies can store and how this information allows us to make accurate inferences, who do you think is spying on you more: NSA, Apple, Google, or Facebook?

When Edward Snowden leaked classified documents that revealed the government's efforts to gather people's personal information,[46] many people were outraged.[47] People in Germany were particularly outraged when they learned that the NSA had monitored Chancellor Angela Merkel's cell phone.[48] One indication of the extent of the outrage is that a leading German newspaper, *Der Spiegel,* has an ongoing special column called "SPIEGEL's Top NSA Reports," which focuses on the NSA's spying activities.[49] Still, few people are aware of the magnitude of corporate spying on the public through people's smartphones.

The hypocrisy of some Internet-based corporations is enormous. Their hypocrisy results in enormous profits. These companies want you to think they are standing up to the government and protecting your privacy, but at the same time they are selling your secrets for huge profits. Their double-dealing deserves to be called out. Please help me spread the word.

Playing with Emotions Is Very Profitable

Manipulating emotions has become a science. The practice now occurs on an industrial level. In 2014, Facebook ran an experiment ("Experimental evidence of massive-scale emotional contagion through social networks")[1] on randomly selected users of its website to see if Facebook could affect outcomes by manipulating users' news feeds. Indeed, Facebook proved that it could manipulate people's emotions and influence outcomes. In the wake of this came a public outcry. The *New York Times* states that Facebook defended itself by citing its user agreement: "The company says users consent to this kind of manipulation when they agree to its terms of service."[2]

That sounds a bit cheeky, doesn't it? Social networking sites' privacy settings and privacy policies change frequently, and sometimes without warning.[3] Facebook insinuates that a user should have realized that he or she signed up to be a guinea pig after reading and consenting to the user agreement. The company's response almost smacks of, "Shame on you if you did not read and understand the user agreement." You are not alone if you were not aware that you agreed to be experimented on or to be subjected to "emotional contagion."

Few people understand user agreements. Lorrie Faith Cranor, associate professor of computer science at Carnegie Mellon University, stated, "We find that the traditional

English-language privacy notice is impenetrable to most people" and "People read a privacy policy, and they can't figure it out."[4]

* * *

Advertisements pitched to the subconscious are more effective than many people are willing to admit. If you want more details, then I encourage you to read *Seducing the Subconscious,*[5] a book written by Robert Heath, PhD, and published in 2012. Many people adamantly refuse to believe they can be influenced or manipulated at a subconscious level. However, this is often possible—and with very dramatic results. "Advertising works best when people think it does not work at all," writes Jean Kilbourne, author of *Deadly Persuasion: Why Women and Girls Must Fight the Addictive Power of Advertising.*[6] Provided that there is just a bit of truth to Jean Kilbourne's book title, we may expect, as advertising becomes more effective with new technologies, advertising in the future to be even more addictive, regardless of the target's gender.

In the past, advertising consisted of predominantly one-way communication. Marketers would design advertising campaigns for a broad audience and then, after launching the campaigns, hope for the best. However, much has changed recently with new technologies. Now, advertising consists of two-way communication, as marketers can tailor a specific advertising campaign to appeal to certain individuals. Marketers may then determine the campaign's effectiveness on individuals without the latter's being aware that they are being monitored. Personal information is very valuable to marketers. With detailed information about individuals and not just audiences as a whole, marketers may develop much more effective advertising campaigns to appeal to individuals' emotions. The more those marketers know about individuals, the easier it is for them to manipulate those individuals' unique emotions.

Do you think it is possible to change the attitudes of many men who prize their masculinity and steadfastness in order to get them to do something they perceive as effeminate? That is exactly what Leo Burnett did with the Marlboro Man advertising campaign. Marlboro is a brand of filtered cigarette that was introduced in 1924 by Philip Morris. It was initially marketed to women with the slogan "Mild as May."[7] Smoking Marlboro cigarettes was considered effeminate. However, some men's resistance to the Marlboro brand was extremely fierce for reasons that went beyond their perceptions of its being effeminate. Those sentiments are not appropriate to discuss here. Instead, it is sufficient for us to understand that the resistance was extremely fierce and that it set the stage for Philip Morris's dramatic turnaround.

After Surgeon General Leroy Edgar Burney warned people in the 1950s that smoking was linked to cancer, Philip Morris (Marlboro's maker) sought to rebrand Marlboro, transforming it from a feminine brand to a masculine brand. Marlboro cigarettes had filters, and some believed the filter would decrease the incidence of cancer caused by cigarette smoking. Still, many advertising agencies turned down Philip Morris's requests for a new campaign because many ad executives thought it would be "very risky" to rebrand Marlboro cigarettes.[8] However, one advertising executive, Leo Burnett, took on the challenge. He ultimately prevailed and did so brilliantly. He changed many men's attitudes by appealing to their desire to appear masculine, but within bounds. He used pictures of men who were "handsome, solitary, tough but with a soft spot for wayward cattle."[9] Marlboro's market share rose from less than one percent to the fourth best-selling brand in one year.[10] Within a few years, Burnett's advertising campaign made Marlboro the most valuable brand in the world[11] because of its clever pitch to the subconscious. Indeed, Harvard Business School Professor Emeritus Gerald Zaltman, PhD, states that the subconscious is an area that, if leveraged by marketers, may secure a competitive advantage.[12]

If you get the point, then skip ahead to the next chapter. However, if you are skeptical of the power of emotional elicitation and subconscious solicitation, or if you just want to know more, then consider how some advertising experts explain the phenomenon of appealing to the subconscious.

- Martin Lindstrom, author of *Brand Sense* (2005) and of the *New York Times* and *Wall Street Journal* best seller and of *USA Today*'s "Pick of the Year," *Buyology* (2010) and *Brandwashed* (2011), respectively, explains that many consumers have been emotionally seduced into parting from their money.

- Marc Andrews and Dr. van Leeuwen, authors of *Hidden Persuasion: 33 Psychological Influences Techniques in Advertising* (2013), provide several examples of how marketers tap into consumers' basic needs in ways that the latter are "unable to fully resist."

- Roger Dooley, author of *Brainfluence* (2012) and publisher of *Neuromarketing*,[13] a popular blog on neuroscience, explains that marketers, if they wish to be successful, should appeal to the prospective buyer's emotions and unconscious needs.[14]

- Jean Kilbourne, advisor to two former surgeons general[15] and internationally recognized for her work on the harmful effects of alcohol and tobacco advertising, shows how marketers inspire consumers' strong feeling of emotional attachment to a product or service. Some of her works include *Killing Us Softly* (2010), *Deadly Persuasion: Advertising & Addiction* (2004),[16] *Slim Hopes: Advertising & the Obsession with Thinness* (1995), and *Pack of Lies: The Advertising of Tobacco* (1992).

- Robert Bishop, chairperson and founder of BBWorld Consulting Services in Geneva, Switzerland, states, "Twenty first century educated consumers have less control over their rational choices and decisions than they may think!"[17]

Manipulating with Fear

Fear is a very powerful emotion.

Edna B. Foa, PhD, and Michael J. Kozak, PhD, experts on the subject of fear, wrote in their 1986 paper "Emotional Processing of Fear" that fear "serves as a program to escape or avoid danger."[1] Fear is essential to our survival. It appears that a fear response is hardwired into our brains and is given very high priority. "Human beings are hardwired to fear things—the lion in the grass, the assailant in the alley—and if one of those fears gets realized, we may never settle down again."[2]

Our brains are hardwired in such a way that enables us to respond very quickly to threats and dangers, which increases our chances of survival. With the exception of daredevils and show-offs, fear leads people to keep a safe distance away from grizzly bears, dangerous-looking people in dark alleys, etc. Fear and the associated fear response are not limited to things that pose a threat to life or things that promise serious bodily injury. Perceptions of danger may induce a fear response. Fear may be elicited by a perceived loss of liberty (this may be why life and liberty are named as inalienable rights in the US Constitution). Fear may also be elicited when one perceives a risk to his or her social status. This type of fear is easy to see in the group dynamics of teenagers as their social groups form, enlarge, and deal with the emergence of new leaders. Fear may also be triggered when a person loses (or perceives that he or she has lost) anything of value. The perceived loss of power or control, or the worry that something valued might be lost, may also induce a fear response.[3]

To understand how the mind handles fear differently than it handles other thoughts and memories, take this simple test. Think back upon your preteenage years and recall all the things that scared or frightened you. That is right—all of them. Take your time, and be honest with yourself. You do not have to tell anyone, and you do not even need to write those scary things down.

This simple self-test will help you understand a very important and very powerful phenomenon. Think of all the things that made you apprehensive, no matter how silly they may seem now, long after they occurred. Choose things that caused you so much fear or anxiety at one point in time in your life that you are able to recall them now, many years later.

Next, think about your ninth birthday celebration. Other than the usual family and friends, think about who else attended your birthday party. Of the usual family and friends, who was missing? Do you recall any gifts or presents you received? Who said what? (By the way, it does not count if you can recall—or guess—what kind of food and dessert you ate, as you likely were served your favorite meal, favorite dessert, or both. After all, it was your birthday.) Most people can recall many of the things that scared them as they were growing up; however, few people can recall the details of any randomly selected birthday, which is probably one of the most joyous and memorable moments of the respective year.

This exercise highlights how our brain prioritizes events related to our survival. It's much more important to keep away from grizzly bears, dangerous people in dark alleys, etc., than it is to remember what should have been one of the most joyous days of any particular year of one's childhood. From a survival perspective, if we can avoid bears and dangerous people in dark alleys, then we can increase the probability of our enjoying another birthday. To put it another way, the mind handles fear with much more intensity and with greater ability to recall details than it handles joyous events.

For most people, the probability of being killed by a bear or by dangerous people in a dark alley is very remote, whereas the probability of having another birthday is very high for most children in developed countries. Sometimes the probability that bad things will happen does not correlate well with the things or situations we fear will actually happen. The mind focuses on the things that scare us, even when those things occur with very low frequency. Keeping those scary things readily available for recall in the mind is often helpful for our survival, even if the probability (frequency) of those things actually occurring is very low. In contrast, the mind does not have the same propensity to recall the details of joyous events, which occur with higher probability (frequency).

It is very important to note that fear of, or the perception of, danger is not always associated with a realistic probability. This is one of the characteristics of posttraumatic stress disorder (PTSD).[4] Understanding this manifestation of fear has greatly aided in the treatment of people who are afflicted with PTSD, but it has also given marketers powerful new tools to use in manipulating customers. For example, televised advertisements of home security companies often dramatize violent home invasions, which lead viewers to imagine their loved ones being suddenly in peril. This technique is preferable to a dramatization of a family returning home to find their jewelry missing and to see that their home had apparently been burgled. The first scenario leads to the sale of more home security systems than does the second scenario, even though the second scenario is, by far, more common than the first.[5] In his books *Risk: The Science and Politics of Fear* (2009) and

Science of Fear: How the Culture of Fear Manipulates People's Brains (2009), Daniel Gardner provides many other examples of how fear is used to manipulate people and to sell products or services.

Although fear is a very powerful means of manipulation, fear appeals are very tricky to use effectively.[6] To be effective, the "dose" of fear must be carefully adjusted for each individual. If the potential customer is overly stimulated by the advertisement, then this may cause him or her to respond in the manner commonly associated with fear: fight, flight, or paralysis. However, when presented with lower "doses" of fear, people often make powerful decisions about how to spend their money. This latter technique marks the sweet spot for marketers. Every person has a different threshold (or tolerance) of fear. The same fear appeal that manipulates one customer into buying a product or service that promises to reduce his or her fear (elicited by an advertisement) may turn off other customers. Therefore, mass advertisements by way of print media or TV often aim for the lowest level of fear so as to manipulate some customers and refrain from overly frightening others.

This is just one area where smartphones and their apps will be superior to other media. Smartphones will be able to monitor an individual's emotional response to fear by using several different technologies (which are discussed in various parts of this book). In this manner, smartphones will be able to provide important clues about the effectiveness of a fear appeal for an individual. In addition, a collective network of smartphones with this technology implanted will help marketers refine and improve the fear appeal so that the advertisement is maximally effective for various segments of the population.

Although this will be very profitable for marketers, the increased effectiveness of fear appeals (with the help of smartphones and apps) may lead to increased generalized anxiety in various segments of the population. Many people are already overly worried, thinking that the world is more dangerous than it is in fact.[7] Many people are increasingly afraid [8] because "fear sells. Fear makes money."[9] Therefore, we can expect to see more generalized anxiety in many people in the near future as smartphones and their apps deliver more effective and more targeted fear appeals.

The "Belongings Test": What You Buy, and Where You Go, Reveals More than You Realize

What belongs to people says a lot about them. Five decades ago, just knowing a few things about a man was enough to make accurate inferences about him. If he preferred a particular style of ink pen, owned a specific style or color of car, smoked a particular brand of cigarettes, and used specific hair and shaving products, then one could accurately infer that the man was a salesperson who wore loud undershorts. This could be accurately inferred without ever seeing the man or his shorts.[1] This sort of thing was done over fifty years ago, when making such a determination was reliant on only a few pieces of information. With thousands or even millions of pieces of information about you combined with much better analytics, guess what sort of inferences can be made about you, including whether your underwear is loud or nondescript? Many more people than you would have imaged know much about your underwear and many more of your possessions and preferences. It is all about making inferences. The more information someone can gather about you, the more accurate the inferences he or she can make about you.

Sam Gosling, PhD, an associate professor of psychology at the University of Texas at Austin, has made a career of studying the correlations between people's possessions and their personalities. In Gosling's 2009 book, he approaches this serious topic from a humorous perspective. *Snoop: What Your Stuff Says about You* shows how expert snoopers can make very accurate inferences about people just by looking at a few of their possessions. Gosling also reveals how body movements reveal much about personality, what characteristics are associated with narcissism, what a single picture of a car says about the driver, etc. The list of inferences is long and fascinating.

Snoop: What Your Stuff Says about You reveals how much can be inferred about a person just by understanding a few things about him or her. Just as smartphones are an improvement on Allen Turing's bombe, the methods of inferring used for spying during WWII have also been greatly improved upon. In his book, Sam Gosling explains how the Office of Strategic Services developed the "Belongings Test" to determine which candidates would make good spies.[2] During this test, "the candidate was taken to a bedroom and given four minutes to study 26 items placed around the room, such as clothing, written materials, a ticket receipt, etc. Next, the spy candidates completed a 36-item questionnaire designed to measure their ability to observe and draw conclusions about the owner of the belongings."[3]

Do you see any similarities? Alan Turing's bombe helped extract secret information from Enigma messages during WWII. The Belongings Test was developed during the WWII era to determine who would be a good spy by making accurate inferences about candidates just by someone's walking around their bedrooms for four minutes. As Alan Turing's bombe has led to the development of the smartphone, the Belongings Test has led to algorithms that can make very accurate inferences about people after evaluating just a few items. Your belongings provide information that can be assembled together to provide a very detailed psychological dossier on you.

Think about how the situation has changed drastically. Techniques developed to select and train spies whose efforts would protect us from our enemies have now been used to develop algorithms to spy on us. The information gleaned about us is sold to data brokers and data analyzers, who then make very accurate inferences about us. Data analyzers, in turn, sell the inferences to marketers. Marketers, in turn, use the very reliable methods they have developed to manipulate emotions or to turn on the brain's reward circuit in a manner similar to cocaine.

These days, data analyzers can provide very accurate insights about you, especially when they link this information with other information about you gained through your smartphone and its various apps. Even museums are getting into mining "Big Data" on visitors. This raises the question of whether visitors are viewing the art or the art is viewing the visitors.[4]

In 2010, Google's CEO Eric Schmidt was quoted as saying the following about Google's users: "We know where you are. We know where you've been. We can more or less know what you're thinking about."[5] That was five years before the publication of this book. Google prides itself on innovation, which leads me to believe that Google has likely improved its data-mining abilities significantly in the past few years.

Retailers realize the value of matching purchases to people and then making inferences from that data. Credit card companies also have access to that data, which they also use to make inferences. However, credit card companies are reluctant to share or sell the data or the inferences made from it. Credit card companies make their money by charging interest rates and financing fees, not by selling information. If credit card holders found out that a credit card company sold data about them, then they would probably not use that company's credit card, figuring that the company had sold them out. This is a cause of concern for credit card companies.

Alternatively, there are many "reward" programs offered by retailers. Retailers offer discounts for customers who participate in a program that matches their purchases with their name or telephone number. By analyzing this data, retailers can make strong inferences about customers and thereby increase sales. On February 16, 2012, the *New York Times* ran an article titled "How Companies Learn Your Secrets," which provides insight into how companies learn your secrets from what you buy.[6] One retailer was able to tell if a woman was pregnant even when she did *not* want others to know.[7]

Because much can be inferred about you from your purchases, some smartphones are starting to circumvent the credit card companies in determining how, when, and where you spend your money. Google has started an e-commerce service, but it is unlikely to be profitable. Google is likely using the service to gather more information about customers.[8] Facebook recently introduced a payments system for its Messenger service.[9] This will provide Facebook with more information to use in making accurate inferences about its users. On January 29, 2013, Apple filed a patent application for "Method and System for Managing Credits via a Mobile Device."[10] This system, called iMoney[11] or Apple Pay,[12] will give Apple unprecedented insight into users' purchases, which will enable the company to make many accurate inferences. If you are a female, then Apple will soon be able to infer if you are pregnant even if you do not want anyone to know. Whether you are a male or female, Apple may soon be able to infer a lot more about you than you want anyone to know. If you want more information about this, then you can read and draw your own conclusions from a *Huffington Post* article of March 19,

2015, titled "5 Extremely Private Things Your iPhone Knows about You."

* * *

Where you go and how you get there reveals much about you. Many people were alarmed to learn that the CIA helped US marshals implement technology that can pinpoint someone's location with a margin of error of ten feet.[13] However, the US government's efforts likely pale in comparison to corporations' ability to gather location and travel information on people. Many corporations are scrambling to get more information about customers.[14] As marketers seek more location information about customers, large telephone companies become more willing to sell huge amounts of data indicating their customers' locations and travel habits.[15] Snapchat, a photo messaging service, has been overzealous in gathering location data and is now under a twenty-year decree from the Federal Trade Commission to cease the practice.[16]

In 2014, a bill was introduced in the Senate that would have provided the customer with more privacy and control over his or her location information.[17] The bill was not enacted, probably because this information is very valuable to marketers and because most people do not realize they are being tracked by many corporations almost as if they were under round-the-clock surveillance by the FBI or the CIA. Many smartphone apps covertly report your location data[18] and corporation have purchased location data from cell phone companies.[19] Perhaps even more location data is available from a new transportation service.

Uber fares are lower than taxi fares in most cities except for New York and Philadelphia;[20] however, there may be more to the Uber story. When and where customers are picked up reveals much about them. When and where they are dropped off also reveals much about them. How long they stay before being picked up adds even more information. Uber hired neuroscientists to analyze the data it accumulated.[21] The neuroscientists demonstrated that Uber's data is ripe for making inferences about Uber's customers. Apparently, venture capitalists were very impressed; as a result, a lot of money flowed into Uber. At least one expert suggests that Uber sells its customers' location information.[22] For legal reasons, I am not alleging that Uber was or is selling customers' private information. Instead, I will provide an assessment of Uber's capability so that you can make your own inferences about whether Uber and other corporations would pass up huge profits in order to protect your privacy.

Google, one the world's best analysts of massive amounts of data, took an interest in Uber and invested $258 million in the company in 2013.[23] Uber's value soon skyrocketed. It was estimated to be $41.2 billion in February 2015.[24] Perhaps Uber's ability to glean inferences about its customers added significantly to the company's market valuation. The high valuation is extraordinary and unlikely to be explained by any other factor than the value of customers' information gathered to make inferences.

Apple, maker of the famous iPod, iPhone, etc., recently announced that it wants to build electric cars.[25] Google is working on producing driverless cars[26] and is rumored to be working on designing electric cars.[27] If you were amazed by Uber's astronomical rise in market valuation and you struggle to understand why it occurred, then all you need to do is follow the money. Given Uber's potential to gather vast amounts of information to make inferences about customers, do you think it is a coincidence that two information technology companies—Apple and Google—have also recently decided to develop electric cars and driverless cars? Google also recently announced that it is developing delivery drones.[28] By inserting itself into the logistics chain, Google will be able to learn more about what people buy and from whom. This would provide Google with more information to make inferences about customers.

Corporations are constantly trying to come up with new ways to gather information about consumers.[29] What started out as the Belongings Test, in which spy candidates had four minutes to study twenty-six items in order to glean insights about the occupant of a room, has evolved into a multibillion-dollar industry.

Big Data Gets to Know You, One Very Small Piece of Information at a Time

New technologies can take thousands—even millions—of small, seemingly disconnected pieces of information about a person and assemble them into a psychological profile or psychological dossier (if you prefer the latter term). In many cases, these psychological profiles are very accurate. These profiles include things that are deeply hidden in people's minds. How did a concept that was merely the quarry of science fiction a few years ago become something that is now nonfictitious?

Recall Frankenstein's comment, "I longed to discover the motives and feelings of these lovely creatures." What Frankenstein longed to do can now be done on an industrial level. Frankenstein wanted to get to know the villagers so he could befriend them. Modern-day equivalents want to get to know you, but they do not want to befriend you. They just want to know you very well in order to make money from you.

The term *profiling* is widely used, but it is subject to drastically different legal interpretations. Racial profiling is illegal for law enforcement officers, but psychological profiling is not illegal for marketers. Psychological profiles or dossiers are huge moneymakers. If you want to know more about constructing a psychological profile or a psychological dossier, then read the more detailed explanation included in this book's notes.[1]

* * *

In chapter 14 of Mary Shelley's novel, Frankenstein takes Victor's papers without the latter's permission or knowledge. More important is that Frankenstein reads Victor's papers and makes inferences from them. The pertinent passage of Shelley's novel is found in the notes at the end of this book.[2]

By 2004, Google was using computers to read its customers' e-mails and to make inferences about those people. On page 12 of an amendment to a registration statement filed with the Securities and Exchange Commission in 2004, Google states, "Gmail uses computers to match advertisements to the content of a user's email message."[3] A few years later, a class action lawsuit was filed against Google for reading people's e-mail (Gmail). Google attempted to have the case dismissed.[4] The case was allowed to go forward, but not as a class action lawsuit.[5] Google claims that users should have understood that they agreed to have their Gmail read, as some wording in Gmail's terms of service indicated as much. The statements to which Google refers, however, are anything but clear. The tortuously worded statements are included in the notes at the end of this book. You may read them and try to make sense of the convoluted legalese.[6]

Google tried to convince Judge Lucy H. Koh that the statements made in Gmail's terms of service were enough to explain to users that Google was reading their Gmail. However, Judge Koh "rejected Google's contention that all e-mail users, regardless of whether they had viewed any disclosures, had impliedly consented to the alleged interceptions."[7] Google also came under fire for reading Gmails of children who used Google's Apps for Education product, which is used by more than thirty million people. As a result, Google announced that it would "revise some of the policies governing Google's 'Apps for Education' products."[8] Whether Google would destroy the information it had gathered on children or would continue to use it was not explained.

Because of Judge Koh's ruling in March 2014, Google changed its user agreement in April 2014 so that Gmail users who read the agreement would clearly understand that their Gmail would be "scanned."[9] Google defended its practice of scanning e-mails by claiming that the scanning is an automated process and by saying that its employees do not read the e-mails.[10] However, Google failed to mention that the scanning would be done by computers, which likely are "smarter" than people.

The foregoing sounds a bit Frankensteinish, doesn't it? Instead of a bumbling oaf like Frankenstein reading your Gmail, modern computers read the e-mails more thoroughly than humans are able to do.

Consider Watson, an artificial intelligence computer built by IBM that can read and understand natural language.[11] In 2011, IBM's Watson competed against people who had previously won on the game show *Jeopardy!* For the feat of prevailing over the two most recent human competitors whose *Jeopardy!* scores were highest, IBM won the first-place prize of $1 million.[12] I am not alleging that IBM's Watson is reading your e-mail. However, Watson revealed computers' capability to communicate using natural-based language. Now, in 2015,

Google probably has computers that are better than the 2011 version of Watson. IBM's Watson 2015 is twenty-four times more powerful than Watson 2011.[13]

Google wants you to believe that it is protecting you from government invasion of your privacy.[14] In addition, Google seeks restrictions on the government's ability to read your e-mail.[15] However, Google has been reading e-mails—or, to use their words, "match[ing] relevant ads to the content of email messages"—since before 2004.[16] Google has been developing artificial intelligence for several years,[17] yet the company is very secretive about how it will use this artificial intelligence. Is Google developing artificial intelligence so that its supercomputers can read your Gmail like Frankenstein read Victor's papers? If so, then Google's claim that it is protecting your privacy (because humans do not read your Gmail) sounds like double-dealing.

Many Internet-based corporations are much more secretive than is the US government. Whereas civil libertarians are able to force the government to give up some of its secretive ways, corporations may hide behind the veil of "proprietary information," claiming that secrecy is necessary to protect them from competitors. In a March 2014 article titled "Google Wants E-Mail Scanning Information Blocked," *Bloomberg Business* reported that Google wanted to redact information on how it mines data from personal e-mails.[18]

Did you catch the contradictions in that quotation? Google wants to have full access to your Gmail so that its computers can read your e-mails. At the same time, Google wants you to think that it is trying to protect your e-mail from government intrusion. However, Google does not want you to know they extract your information from your Gmails.[19] Google is not the only corporation that collects information from e-mails that a person has not yet sent or has decided not to send. One tech consultant claims that Facebook "DOES collect the text you decided against posting ... despite the company's claims to the contrary."[20]

* * *

In 2002, Steven Spielberg produced the science fiction movie *Minority Report,* starring Tom Cruise. One of the themes of this movie was precrime.[21] The idea was that crimes could be predicted and therefore prevented if law enforcement officers monitored potential criminals by using futuristic technologies. This science fiction movie is set in 2054. However, several technologies exist today that could make many of the scenarios presented in the film possible. Predictive technology is mature and is being used widely.[22]

Some predictive technology is able to make inferences about people by analyzing data gathered on them without their awareness that such a thing is being done. Some entities may gather information about you by using recently patented technology that can read your mind. The term *mind reader* makes some people uncomfortable; therefore, these entities use phrases like "accurately reflect a person's thoughts" instead.

The important difference between the science fiction movie and real life today is how the information is used. In Minority Report, the technology was used to reduce crime and protect the public. Today, the technology may be used commercially to provide marketers with crafty new ways to manipulate consumers. A person may be charged more for his or her online purchases than others are if algorithms target the individual as a person who is able and/or willing to pay a higher price.[23] The Wall Street Journal reported that owners of Macintosh computers were quoted higher prices for hotel rooms than were those people who used other brands of computers.[24]

"Anonymous," but Very Well-Known

Because of the increasing public pressure to safeguard people's privacy, many corporations have made what appear to be solid pledges to safeguard their customers' privacy. Still, advertisements targeted to a specific individual are delivered to that person's e-mail account or smartphone. Some people receive new advertisements within minutes of purchasing something either online or from a retail store. Others get advertisements within minutes of performing online searches.

How can your privacy be guarded as well as is promised by many Internet-based companies and other retailers when people receive accurately targeted advertisements very quickly after making a purchase? Your purchases aren't the only things that certain companies are aware of, however. Your medical information is also made available to certain entities. Answer this riddle: how can your medical records be protected by the Health Insurance Portability and Accountability Act (HIPAA), which stipulates that leaking HIPAA information will result in huge fines (millions of dollars) for the transgressors,[1] if some information contained in your medical records is for sale to be used by Big Data?[2] To answer this riddle, all we need to do is follow the money. Medical information is very profitable. Medical information is sold to marketers that sell drugs, devices, and services. Sick and hurting people are often willing to pay for anything that offers even a remote chance of relief. Therefore, the titans of the Internet have found ways to get around privacy laws and HIPPA so they can profit from people's private medical information.

Many corporations promise to guard your information and promise to share only "anonymized" information about you. The concept of anonymized information may sound impressive. When you examine these claims closely, however, you see that they are deceptive. For example, you are so well known that you can watch the same show on cable TV as your neighbor does, but each of you will be shown different advertisements based on your online activity.[3] You are just as effectively targeted for advertisements on your smartphone as you are on broadband cable, and it is all done with anonymized information.

In everyday usage, the word *anonymous* implies that individuals are unknown or indistinguishable. In the cyberworld, this is simply not true. By not using people's names and Social Security numbers (SSNs), many Internet-based corporations have cleverly made a mockery of privacy laws and of the phrase "personally identifiable information" (PII). A Federal Trade Commission staff report of 2010 "acknowledged that the traditional distinction between the two categories of data [PII and non-PII] has eroded and that information practices and restrictions that rely on this distinction are losing their relevance."[4] Still, most people believe that they cannot be identified if they do not provide their name or SSN. That is false, as the opposite is true. It may seem counterintuitive that information technology can identify people much more readily *without* having access to their names or SSNs than it could if it did have access to that particular information.

* * *

Facebook does not allow its advertisers to have direct access to its users' personal information. The social networking site will accept information about a user from other sources in order to match that user's personal information to certain advertisements, although Facebook claims that the user is not identified because his or her information has been anonymized[5] in a process called hashing.[6]

To understand hashing, we need to go back to Alan Turing, who worked in the area of British cryptanalysis (code breaking) in WWII, and then work forward to the solving of a problem faced by casinos. Casinos wanted to identify those people who did not play by the rules. Casinos identify people who are not playing by the rules by using hash functions[7] or the more specific cryptographic hash function.

Here is an illustration of hashing,[8] one that is intended to show how hashing can blur the lines between cheating and not cheating. (Note that this is not how Facebook is using hashing.) Assume the following: Tom, Sam, Sally, Mary, and Joe are taking a class in mathematics that includes take-home examinations. In indicating their consent to a user agreement, all of the students, including Tom, Sally, and Joe, agree not to share their answers. However, Tom, Sally, and Joe come up with a plan to help each other on the exam without sharing the actual answers. The five students decide that, instead of sharing the actual numbers, they will encode their answers. This way, they think they can deny sharing the actual numbers if they are caught. After all, they reason, they are not sharing the numbers; rather, they are just sharing encoded answers, which do not actually represent the numerals that are the solutions to the test questions. After searching the Internet to find a hash generator, they find one they like: the MD5 Generator.[9]

Once one person completes a question and obtains what he or she assumes is the correct answer, he or she encodes the answer and then shares that encoded answer with the other four students. All five do the same, and then they all compare their encoded answers. If the answers are all the same, then the five have increased confidence that the answer is correct.

Let's say the first question requires a series of calculations. Tom, Sam, Sally, Mary, and Joe all come up with 247 as the answer. Following are the coded messages that all five would give to each other:

    Tom: 3cec07e9ba5f5bb252d13f5f431e4bbb
    Sam: 3cec07e9ba5f5bb252d13f5f431e4bbb
    Sally: 3cec07e9ba5f5bb252d13f5f431e4bbb
    Mary: 3cec07e9ba5f5bb252d13f5f431e4bbb
    Joe: 3cec07e9ba5f5bb252d13f5f431e4bbb

Since all five hashed answers are the same, Tom, Sam, Sally, Mary, and Joe are reasonably confident they have all solved the problem correctly. They move on to the next problem.

Let's say, however, that Joe got 351 for his answer; therefore, his hashed number looks different from the other four.

    Tom: 3cec07e9ba5f5bb252d13f5f431e4bbb
    Sam: 3cec07e9ba5f5bb252d13f5f431e4bbb
    Sally: 3cec07e9ba5f5bb252d13f5f431e4bbb
    Mary: 3cec07e9ba5f5bb252d13f5f431e4bbb
    Joe: efe937780e95574250dabe07151bdc23

Joe goes back, reworks the problem, and gets an answer of 247. He hashes his answer and sends it to Tom, Sam, Sally, and Mary. Since all five answers are now the same, Tom, Sam, Sally, Mary, and Joe are reasonably confident that they have all solved the problem correctly, so they go on to the next problem. Although all five could have solved the problem wrong, it is less likely that all five would have gotten the same wrong answer through this process. Instead, it is more likely that all five got the correct answer.

Tom, Sam, Sally, Mary, and Joe complete the remainder of their take-home exam and turn in their answers. All five get 100 percent correct. However, they are surprised when the teacher accuses them of cheating. They adamantly deny they cheated, saying they did not share the answers.

The teacher refers them to the academic integrity resolution board. In front of the board, Tom, Sam, Sally, Mary, and Joe again adamantly deny they cheated, saying they did not share the *numbers* that were the answers to the questions. However, the teacher, having seen students use hashing before, was prepared. She had intercepted the five students' coded/hashed answers. She had permission to do this because all five had clicked "Yes," indicating their agreement to abide by the terms set out in the twelve-page-long user agreement, which included a statement about forfeiting privacy: "We voluntarily give up our expectation of privacy in academic matters pertinent to this class."

The teacher shows the academic integrity resolution board all of the intercepted messages between the five, which show that the hashed answers are merely interchangeable substitutes for the numerical values. To add more proof to the allegation of cheating, the teacher shows the board how the same algorithm that encoded the numbers to hashings can be used to decode all the hashings and convert them back to numbers. Unsurprisingly, all of the decoded hashes match the numbers that are the answers to the exam questions.

Imagine that you are one of the members of the academic integrity resolution board. You must decide if Tom, Sam, Sally,

Mary, and Joe cheated. However, before you decide on this notional scenario involving only answers to a test, consider the situation from a different viewpoint. Hashing as it is used by Facebook and others is more complicated than this example of students cheating. However, this example provides a framework to understand how information can be transformed and still be usable. What if it were not just numbers on a test but your personal information, including your mood data as well as information gathered from voice analytics, facial action coding, eye movements, and brain waves? Moreover, what if the results were being sold in an allegedly anonymized format?

This is another case of how methods that were developed by spies to protect a nation from its enemies are now used by corporations to spy on people and sell their information to others for the purpose of commercial gain. Many experts claim that Facebook's efforts to anonymize its users' information are ineffective. Several experts reject Facebook's practice as a viable method of protecting users' personal information.[10] Additionally, the Federal Trade Commission also believes that the social network's hashing of user data is overrated.[11]

The example of the five students cheating on a test is used to illustrate how encrypting and decrypting data can blur the lines between cheating and not cheating. Commercial encrypting and decrypting often involves one party sending information to a second party. In this next example, I use a notional person with a notional SSN and credit card information. John Doe, Social Security Number 123-45-6789, has a credit card with a number of 9876-5432-1987-6543. Encoded with an MD5 hash generator, his name is converted to "4c2a904bafba06591225113ad17b5cec," his SSN (without the dashes) is converted to "25f9e794323b453885f5181f1b624d0b," and his credit card number (without the dashes) is converted to "f7f1118e4478ac1263c01ff26f2a6474."

Since the public is aware of the MD5 hash generator, John would have to use a slightly different method of encrypting his information. If John used strong encryption keys, then he could conduct business on the Internet with his name, SSN, and credit card number encoded. He could then be reasonably sure that thieves would not able to use his information if they intercepted it. Google, Facebook, and other companies encrypt your information and tell you that your information is protected and anonymized. However, you can see that if your credit card number can be routinely encrypted and decrypted every time you use it, then so can all of your other information. Encrypting and decrypting information gleaned during commercial transactions is a fascinating topic, but a full discussion of this practice is beyond the scope of this book. If you would like more information, then I recommend that you read Simon Singh's *The Code Book*.[12]

* * *

Some people want to try to beat the odds of casino games by using dubious methods. Some would-be cheaters may try to use several different identities in order to make it appear that a single person is not making a lot of money. Likewise, a dealer at a blackjack table may collaborate with a roommate, neighbor, relative, or someone else who offers to split the winnings with the dishonest dealer.

To find out if one person is using different identities or if the dealer and the winner are roommates, neighbors, etc., a software engineer, Jeff Jonas, designed a system in the 1990s[13] to catch would-be cheaters. The system used hash functions in a different way from what is described above in order to resolve questions of identity[14] by finding duplicate or similar records—utility records, driver's license records, vehicle registration records, etc.—that exist within various electronic systems.[15] This system worked very well for the casinos.

Jeff Jonas, after selling the system he had created along with his company to IBM in 2005, is now the chief scientist of the IBM Entity Analytics group.[16] IBM (and probably Jeff) has since added an additional feature to the original system—anonymous resolution.[17] The software can quickly "unanonymize" the information that companies claim is anonymous. This conflicts with Google's claim of storing users' anonymous information forever. It also conflicts with Google's claim of using an effective "anonymous identifier"[18] to protect users' identity and information. These claims are revealed as deceptive when you understand how casinos can protect themselves from people who try very hard to be anonymous and to cheat the casinos.

Google claims to protect user information by using another method to anonymize that information. Your computer's address is known as an Internet protocol address, or IP address. IP addresses include only numerals (not letters). Initially, they were only 32 digits long. However, with the rapid growth of the Internet, longer IP addresses were assigned to computers in order to accommodate the increased number of addresses. IP addresses may now be 128 digits long.[19] Google, a search-engine company, anonymizes user information by dropping the last part of a computer's IP address.[20] However, it drops only enough of the IP address to make it different from 254 other IP addresses.[21] With other information about a user (e.g., a telephone number, an e-mail address, etc.), Google can quickly unanonymize a user's information by matching it against that of 254 other people, instead of 7 billion, and in this way, detect individual characteristics and attributes. Remember Jeff Jonas's software, which made it easy for a casino to spot a dealer's roommates who were trying to remain anonymous while cheating the casino? Matching information to one user by comparing it to the information of 254 other people is so simple that computers can do it in milliseconds.

* * *

You have two identities. Let's first discuss the identity that is associated with your name, address, Social Security Number, etc. These things are part of your personally identifiable information. One's PII[22] is claimed to be protected when one is active online. However, that is false. Laws and regulations regulating how PII is used are intended to prevent monetary fraud and theft. They are not intended to prevent one's personal information from being gathered and analyzed.

Several laws to protect PII were proposed at the federal level. However, no federal laws have been passed that specify exactly what PII is and what it is not. PII has had a very troubled legal life and is now almost technologically obsolete.[23]

While governments and consumers are focused on keeping people's names and SSNs private, your name and SSN have already become obsolete in the electronic realm. Big Data has found cleverer ways to identify a user instantaneously and separate him or her from everyone else in the world—and with fewer mistakes than if it were using your name or SSN. Your name and SSN present many problems for Big Data. For example, say that Big Data comes across the name Kathy Smith. Is this the twenty-one-year-old Kathy Smith[24] who lives in New York City, or is it the sixty-four-year-old Kathy Smith[25] who lives in Seattle, Washington? Alternatively, is Mike Miller a fifty-four-year-old in Phoenix, Arizona, or a twenty-five-year-old in Denver, Colorado? Adding to Big Data's problem of discerning individual identities is the fact that millions of US citizens have multiple Social Security Numbers.[26] Some people have the same SSN as others.[27] These problems are largely the result of bureaucratic blunders. The many problems with SSNs make it difficult to keep track of people by using their SSNs.

Let's now discuss your second identity, which is your cyberidentity, or digital identity.[28] Big Data has assigned you a digital identifier, which you may not be aware of, that cannot be turned off or deleted.[29] Data gatherers may place tiny files such as beacons on your computer without your realizing it. These tiny bits of code reveal a lot of information about you without identifying you by name.[30]

You may have more than one cyberidentity, depending on which system or systems are tracking you.[31] The notion of a cyberidentity has very quickly and very quietly morphed from an aspect of science fiction to something that has economic value,[32] is readily marketable, and is becoming more valuable to marketers every day. Cyberidentity has given rise to the phrase "personal data economy."[33] If you are interested in more details about how transparent your anonymous data really is, then I recommend you read chapter 3, "Analyzing Our Data," of Bruce Schneier's Data and Goliath: The Hidden Battle to Collect Your Data and Control You (2015).[34]

The Genius of Facebook

Many people attribute rock-star status to Mr. Mark Zuckerberg for his having created the Facebook phenomenon. After all, his face is pictured on the covers of many magazines and alongside many news stories. However, the founding of Facebook is mired in controversy.[1]

While he was attending Harvard University, Mr. Zuckerberg hacked into Harvard's computer services department's computer to obtain photographs of students' faces, which he was not authorized to do.[2] He also hacked two *Harvard Crimson* reporters' e-mail accounts.[3] The reporters were looking into a dispute between Zuckerberg and some other people concerning whose idea it was to start HarvardConnection, which was a forerunner of Facebook.[4] Mr. Zuckerberg was sued by Cameron and Tyler Winklevoss. The case was settled out of court. The Winklevoss brothers were awarded Facebook shares worth $300 million.[5]

Mr. Mark Zuckerberg appears to enjoy being aggressive. Consider one of his statements, "Move fast and break things. Unless you are breaking stuff, you are not moving fast enough."[6] The Belgium Privacy Protection Commission (CPVP/CBPL), which is working with German, Dutch, French, and Spanish counterparts, accused Facebook of trampling over privacy laws.[7] Belgium researchers found that Facebook violated its terms of service by placing cookies—a common way of tracking people's browsing habits on the Web—in *some* people's browsers, even if those people had never visited Facebook.com to sign up for an account. Facebook blamed the problem on a "bug."[8] Really? A computer bug allowed Facebook to accidentally gather more of its users' private information? Is Facebook's software autonomous and out of control? If so, then what other surprises can we expect?

Facebook has conflicted with the Federal Trade Commission (FTC) on other issues and is now subject to a twenty-year user-consent decree imposed by the FTC.[9] However, investors and corporations that are in the business of leveraging users' personally identifiable information were willing to look past these "irregularities" because Facebook quickly turned into a multibillion-dollar source of income. In their rush to support Zuckerberg, some claimed that he was a genius. After all, it would take a genius to turn a social media site into something that could generate billions of dollars. Indeed, there was a genius involved. However, the real genius was born in 1901, more than eight decades before Mr. Zuckerberg was born.

Since Mr. Zuckerberg had few qualms about hacking into reporters' e-mails and into off-limits Harvard computers containing students' photographs, it should be no surprise that he has no qualms about selling information that has been harvested from users' Facebook accounts by some very sophisticated mathematical tools. These mathematical tools were developed in the 1930s.

Few people have heard of Paul Lazarsfeld, PhD, yet he is the real genius behind the Facebook phenomenon, for two reasons. He pioneered modern advertising and he cofounded[10] what sounds like a very odd mixture—mathematical sociology. Without his work, there would be no social media as we know it today, since the results of using mathematical sociology provide the revenue that makes social networking services "free" to users. The service is "free" to users because marketers benefit from the very accurate inferences that may be gleaned by analyzing a user's online behavior and posts. Facebook sells these inferences to its advertisers.

To provide more background on Paul F. Lazarsfeld, I'll discuss a Viennese laundry service in existence in 1930,[11] when Lazarsfeld was a psychology instructor at Vienna University.[12] The owner of the laundry service was struggling and did not understand why Viennese homemakers would not use laundry services. By conducting numerous interviews, Lazarsfeld learned that Viennese homemakers thought it was a sign of failure to use commercial laundry services. These homemakers thought that they would be considered lazy if they did not do their own laundry. Using commercial laundry services was viewed as acceptable only under rare and overwhelming circumstances, such as when there was a death in the family.

Armed with this insight from potential laundry customers, Lazarsfeld suggested ways to overcome their objections. After implementing Lazarsfeld's recommendations, the laundry owner saw his business flourish. Although Lazarsfeld earned a doctorate in mathematics, with a dissertation that covered certain mathematical aspects of Einstein's theory of gravity,[13] he would eventually become known for bringing the "psychology approach" to the study of consumer behavior.[14]

In 1934, Lazarsfeld[15] published "The Psychological Aspects of Market Research" in the *Harvard Business Review*.[16] Lazarsfeld also cofounded the field of mathematical sociology, as mentioned previously.[17] A former president of the American Sociological Association, James S. Coleman, PhD, wrote a book in 1964 titled *Introduction to Mathematical Sociology*. Coleman dedicated the book to Paul Lazarsfeld.[18] As someone who worked with horses and avoided a few near misses from ill-tempered horses, I find it fascinating that James Coleman, PhD, developed some of his work by using statistical methods that were originally used to analyze the number of soldiers in the Prussian army who were killed by horse kicks.[19] The statistical methods used to predict the rate of those occurrences now provide several multibillion-dollar revenue streams for several corporations.

Mathematical sociology has advanced to become part of several different disciplines, including social network analysis.[20] It is now a powerful way to mine vast amounts of information from social network websites. The technical basis for harvesting vast amounts of information from social networking sites was developed long before Zuckerberg was born. All that was needed for Zuckerberg to succeed in this area was for the price of computers to drop and for someone to develop a database connecting large groups of people.

As described in Mary Shelley's novel, Victor stitched Frankenstein together by using various pieces and parts of living things. Computers and algorithms based on mathematical sociology and social network analysis may electronically stitch together very detailed dossiers on users by linking together thousands of seemly insignificant pieces information gathered from personal social media accounts.

Given the freedom to associate with others, people tend to develop relationships with others who share their values, feelings, interests, thoughts, outlooks, etc. Therefore, learning about a person's friends reveals much about the person. A person's friends and neighbors indicate a lot about the person and show how the person is likely to behave.[21] However, Facebook and other social media sites do more than just establish social networks. Everything a user posts on Facebook (comments, photographs, previously uncopyrighted poems, or lyrics to songs, etc.) is the property of Facebook ("you grant us a non-exclusive, transferable, sub-licensable, royalty-free, worldwide license to use any IP content that you post on or in connection with Facebook").[22] Facebook may data-mine everything that a person posts in order to find clues about the person and to use those clues to make inferences. Facebook even has software that can track a person's cursor's location on his or her computer screen.[23]

In a study of 58,466 volunteers from the United States, with an average of 170 Facebook "likes" per volunteer, researchers accurately predicted the user's ethnicity in 95 percent of cases, sexual orientation in 88 percent of cases, political affiliation in 85 percent of cases, and many other personal attributes.[24] Using only a user's 170 "likes," the researchers were able to infer much about the volunteers. The more information Facebook has about a person, the more accurate the inferences will be.

- According to a Senate report, some private companies "collect, mine and sell as many as 75,000 individual data points on each consumer."[25] Using seventy-five thousand data points certainly allows for more accurate predictions based on users' highly sensitive personal data. The Federal Trade Commission estimated that the data broker industry, which is lightly regulated, uses hundreds of billions of pieces of information to make detailed inferences[26] about individuals. FTC Chairwoman Edith Ramirez states, "It's time to bring transparency and accountability to bear on this industry on behalf of consumers, many of whom are unaware that data brokers even exist."[27] A few highlights of the FTC's May 27, 2014, report include the following:[28]

    o "Data brokers combine and analyze data about consumers to make inferences about them, including potentially sensitive inferences such as those related to ethnicity, income, religion, political leanings, age, and health conditions. Potentially sensitive categories from the study are 'Urban Scramble' and 'Mobile

Mixers,' both of which include a high concentration of Latinos and African-Americans with low incomes. The category 'Rural Everlasting' includes single men and women over age 66 with 'low educational attainment and low net worths.' Other potentially sensitive categories include health-related topics or conditions, such as pregnancy, diabetes, and high cholesterol."

o "Data brokers collect consumer data from extensive online and offline sources, largely without consumers' knowledge, ranging from consumer purchase data, social media activity, warranty registrations, magazine subscriptions, religious and political affiliations, and other details of consumers' everyday lives."

How do data brokers make these inferences? They do it by taking millions or billions of small pieces of information about a specific person gleaned from social media and then, together with software likely developed with assistance from psychologists and anthropologists, piece together a very detailed psychological profile in a manner similar to reconstructing a person's entire DNA by mathematically reconstructing millions of small segments of his or her genetic code.

Mathematical sociology uses techniques that are similar to those used in epidemiology.[29] Perhaps one of the most powerful uses of mathematical sociology is linking an individual with a group of other people by analyzing the person's social network.[30] Even if a person does not post anything on the social networking site being studied, marketers can discover a tremendous amount of information about that individual by knowing what the person's friends say about him or her. Therefore, when you "friend" one of your friends on Facebook, you are selling your friend out and getting little or nothing for it. Similarly, when one of your friends "friends" you on Facebook, he or she is selling you out and getting little or nothing for it.

The psychological dossiers derived from using the tools of mathematical sociology can very accurately portray a person's emotions, desires, and fears. This is rich fodder for marketers who seek to use this information in order to tailor a message for, and then target that message to, a specific person and thereby appeal to him or her on a subconscious level. This is why data analyzers and marketers are willing to pay billions of dollars for tens of billions of small pieces of information about people.

The increased capability of computers, along with the falling prices of computing power, was a fortuitous thing for social media corporations and was likely the reason for their astonishingly rapid capitalization. With social networks that appeal to individuals as forums for communicating and sharing ideas, schedules, etc., also came the opportunity to harvest huge amounts of information and analyze it with computers and mathematics for the purpose of selling that information to marketers. This resulted in astronomical profits for social media and social network corporations.

Social media may be a current craze, but it is causing problems. Dr. Elian Fink and Dr. Miranda Wolpert, writing in the *Journal of Adolescent Health,* reported, "Sexualised images of women in advertising and social media are leading to an increase in emotional problems among young girls. ... The number of schoolgirls likely to suffer emotional problems also rose from 13 per cent in the 2009 study to 20 per cent—one in five—in 2014."[31]

The Eyes Are the Window to the Mind

The cover of the August 2007 edition of *Scientific American* states, "Windows of the Mind: How Tiny Twitches Preserve Vision and Reveal Thoughts."[1] With the help of clever algorithms, new technology can now shine a powerful searchlight into the subconscious mind. This powerful searchlight may illuminate thoughts and feelings that are tucked so deep in the recesses of a person's mind that he or she may not be aware of them. If the person is aware of the thoughts and feelings, then he or she may be uncomfortable sharing them with friends and family. Yet new technologies may extract this information from a person's mind so that companies may then sell it to the highest bidder.

Eye tracking is well suited to neuromarketing studies.[2] Although eye tracking may have started with evaluating how best to display products on a shelf,[3] recently it has grown very rapidly into something much bigger and more sophisticated. To better understand what is going on, let's look at what happened recently with infrared technology and its ability to track people's eye movements precisely.

For legal reasons, I am not alleging that corporations are currently using these technologies to read minds. Rather, this chapter is merely meant to provide an assessment of recently patented technology. It remains to be determined how comfortable the public will be when these technologies are used on them.

Your pupils can reveal much about you,[4] as can your eye blinks and eye movements. Grants from the US Navy and US Air Force[5] paid for the development of a technology that resulted in a patent titled "Method and apparatus for eye tracking and monitoring pupil dilation to evaluate cognitive activity."[6] This technology can "determine mental alertness level by monitoring point of gaze, pupillary movement, pupillary response ... and assigning the subject to a score indicating the subject's particular mental alertness level in real time."[7] These devices can evaluate cognitive activity by using cameras capable of recording important eye-tracking data, such as pupil dilation, eye blinks, eye movements, and on what the eye fixates.[8] This technology is not limited to military applications. Now commercially available,[9] it yields an index of cognitive activity (ICA).[10] The ICA is based on changes in pupil dilation that occur as a user interacts with a visual display.[11] A person "performing a task can be monitored unobtrusively by a remote camera capable of recording important eye tracking data such as pupil dilation, eye movements and fixations, and blinks. The eye data can be processed, for example, in real time to provide an estimate of the current level of mental alertness or mental proficiency."[12]

> The ICA demonstrates, information derived from the eyes may be extremely useful. But, obviously, more information is available than just pupil size alone. Would the inclusion of other information such as eye movement or blinking make it possible to detect a wider range of mental alertness levels? The answer is yes and the subject matter described herein may be used to estimate mental alertness or proficiency levels from a set of eye

metrics. For example, metrics based on pupil dilation together with horizontal and vertical position of the point of gaze. Pupil dilation refers to any change in pupillary diameter that occurs in a subject's eye. Point of gaze refers to any point of reference upon which a subject's eye focuses (for example, a visual display) for a measurable period of time.[13]

Many companies now offer eye-tracking technology. Consider some of the claims made on various companies' websites indicating the technology's ability to look into a person's subconscious:

- "Eye tracking can be used as a window into people's subconscious thought processes."[14]

- "EyeTrak uses state-of-the art, infrared eye-tracking technology to accurately measure both conscious and subconscious reactions."[15]

- "Tobii eye tracking accurately measures both conscious and subconscious reactions to stimuli."[16]

- "By measuring subconscious behaviour, Eye Tracking allows you to monitor whether a message has been seen and remembered."[17]

- "Today, however, it is possible for market researchers to get information beyond what consumers say and actually find out how consumers subconsciously feel about what they see."[18]

- "Eye tracking captures shoppers' habitual and subconscious behavior in a natural and unbiased fashion."[19]

- "Pizza Hut and Tobii are experimenting with 'the world's first subconscious menu.'"[20]

If you are interested in finding more examples, then you should search the Internet, entering *eye tracking* and *subconscious* into the search box, which will retrieve the websites of a few companies that make similar claims. When I wrote my earlier book about eye tracking and monitoring the subconscious (*Google Glass Can Read Your Mind,* 2014), I noticed that several companies were boasting about their ability to measure the subconscious by using eye tracking. Since that time, corporations appear to have become squeamish about posting information on their websites that includes the terms *eye tracking, conscious,* and *subconscious*. For example, one of the quotations above—"Tobii eye tracking accurately measures both conscious and subconscious reactions to stimuli"—no longer appears on the Tobii website. I was able to find the sentence on a different website, one belonging to Acuity ETS. Tobii still acknowledges that eye tracking can monitor the subconscious. However, Tobii has reframed eye tracking as a driver safety issue: "With eye tracking, we were able to study actual driving behavior in real-time and in detail, and identify subconscious driving episodes/situations."[21]

Eye tracking technology has been around for many years and has been widely used for purposes of consumer safety and research into consumer behavior. Head-mounted displays fitted to consumers have been used by marketers to learn how consumers look at individual products, product packaging, or products in displays, such as on store shelves. More recently, eye-tracking devices have been used to monitor how people look at Web pages. By using these tools, researchers can refine and improve the visual appeal of Web pages, products, and packaging as well as determine how a company's product stands out on the shelf compared to a competitor's product.

* * *

Attention is a selective process. The conscious mind simply cannot process all of the visual information the eyes provide. Many scientists estimate that 95 percent of all cognition occurs in the subconscious mind.[22] To handle this massive amount of visual information, the mind has developed the neurological equivalent of a triaging system to sort through and prioritize visual stimuli. A small part of your visual field is handled by your conscious mind, and the remainder is handled by your subconscious mind.

Your eyes continuously make tiny jerky movements. By measuring these jerky eye movements is how a computer can read your mind. The French word *saccadé* is used to describe a horse rider's abrupt pull on the reins. In the 1880s, a French ophthalmologist, Louis Émile Javal, used the term to describe the jerky motions of the eye.[23] For many years, the purpose and function of these jerky motions remained a mystery. However, scientists have recently revealed that the jerky eye movements keep a person's subconscious aware of what is happening in his or her peripheral vision. Scientists have also shown how saccades increase visual clarity[24] and can cause some optical illusions.[25]

Research into saccades in recent years has accelerated tremendously.[26] Some experts consider microsaccades, saccades, and fixational saccades to have somewhat different functions, but many similarities.[27] Most experts consider them to be related.[28] If you would like to learn more, read two excellent articles, "Visual attention: The past 25 years" (2011)[29] and "The impact of microsaccades on vision: Towards a unified theory of saccadic function" (2013).[30] Fortunately, you do not have to understand all of the nuances in order to understand how these jerky movements can reveal much about your thoughts. (For purposes of simplicity, *saccade* will be used in this text to indicate all three terms: microsaccades, saccades, and fixational saccades.)

Rather than explaining the phenomenon from a scientific point of view, perhaps it would be better to look at it in the context of common experiences. Saccades are responsible for the optical illusion that makes fans, airplane propellers, and other rotating objects occasionally appear to spin backward.[31] The small quick movement of the eye shifts the image on the retina so the light from the image strikes new rods and cones in the retina. When light strikes a rod or cone in the retina, a chemical reaction occurs which results in an electrical impulse being sent to the brain. Once rods or cones discharge the electric pulse they need time to recharge. If the eyes did not make small movements, then objects in your view would fade.[32] The jerky movements of the eye keep vision sharp.

When light enters the eye, it hits the retina, which contains rods and cones. When light strikes the rods and cones, it causes the discharge of electrical impulses. Once discharged, these travel through the optic nerve and then reach the brain. The brain integrates all of these electrical impulses into something the eyes perceive to be a visual image. Once the rods and cones in the retina discharge, however, they need a little time to recharge in order to prevent retinal fatigue.[33] The process described in this paragraph is only one function of saccades. It is interesting to note that amphibians and reptiles do not have this feature, so these creatures can only see things like flies when they move. That is why a fly is often safe so long as it stays in one spot because the frog cannot see the fly. Once the fly tries to get away, the frog sees the fly and quickly catches the fly.[34]

The second function of these rapid eye movements is more relevant to the topic of this book. Saccades are an integral part of peripheral vision. Have you ever been looking attentively at something directly in front of you and then suddenly caught a glimpse of something out of the corner of your eye? That phenomenon is made possible by saccades.

Most people are not consciously aware of the rapid movements of their own eyes or of the objects within their fields of view beyond what they are already focused on. The subconscious mind's exploration of the peripheral vision is largely involuntary.[35] When referring to visual attention, *covert* does not mean "sneaky." In the case of saccades, *covert attention* means attention that which is "not openly acknowledged or displayed."[36] Covert attention is also called reflexive attention,[37] as the attention shifts reflexively to different areas of the peripheral vision when new objects enter the visual field or else move. These subconscious, exploratory saccades can be very accurately measured and converted into an index of covert attention.[38] Computers can measure these rapid eye movements, as they use algorithms that have been developed to accurately detect saccades.[39]

There is simply too much in your field of view for your conscious mind to make sense of it all, especially since your conscious mind handles only about 5 percent of the cognitive workload of your brain (this leaves 95 percent to be handled by your subconscious). A simplified explanation is that your mind triages what is most important to you. What your mind triages to your subconscious becomes that to which you pay covert attention.

Perhaps an analogy will make saccades easier to understand. Think of your subconscious as a personal assistant who monitors the things going on around you but who will not interrupt you unless something very important arises. For example, if you are in a stadium watching a baseball game and focusing intensely on the pitcher, your conscious mind may not notice your friend waving from the end of the row, trying to get your attention. However, your subconscious is aware that someone is waving and will bring the action to your conscious attention. Likewise, if you are riding a bicycle and trying to catch up with the rest of your group, you may not see a car coming at you from the side. However, your subconscious does see the car, and it informs your conscious mind so as to protect you from a potentially dangerous situation.

Your conscious mind stays focused on a very small part of your visual field. However, your subconscious is very actively monitoring the remainder of the visual field—what is commonly known as your peripheral vision. Given that your eyes can look at only one spot at a time, your conscious mind and your subconscious mind both utilize your eyes (i.e., they share them). Your eyes quickly wander in order to monitor the periphery. Most people are not even aware of their exploratory saccades.[40] Still, it is possible to consciously override[41] this automatic subconscious exploration of peripheral vision.

Algorithms have been developed that can accurately detect saccades.[42] By tracking your saccades, these algorithms can detect what objects are of interest to your conscious mind (i.e., those things to which you pay overt attention) and what objects in your peripheral vision are of interest to your subconscious mind (i.e., those things to which you pay covert attention).[43] Your subconscious is very aware of what is in your peripheral vision, even though you are not gazing directly at any object found there. Thus, your conscious mind can stay focused on the small area that is of immediate interest to you (overt attention) while your subconscious handles the things in your peripheral vision (covert attention).

The rate of saccades changes as your subconscious becomes more interested in things in your peripheral vision (i.e., when something new comes into your peripheral vision, the rate of saccades changes).[44] The direction of saccades is an indicator of either subconscious or covert attention.[45] Shifts in covert attention can be identified by a change in the rate of the saccades (the change is detected by comparing the rate to a baseline for a particular individual).[46] Saccades can also reveal the orientation of covert attention.[47]

Saccades are able to reveal what has your conscious attention. They may also reveal your subconscious thoughts, including what you are trying to resist looking at.[48] For example, if you are trying to lose weight and, in your peripheral vision, you spot your favorite dessert sitting on a table, then your subconscious may try to resist looking at the dessert, just as if you told your personal assistant, "Don't look over there!" If you are prone to drinking too much alcohol, then your saccades might reveal to marketers not only your temptation to drink alcohol but also your favorite drink. If you have an eye for pornography, then your saccades may indicate that as well. Your saccades even indicate what you are trying to resist looking at, as mentioned above.[49] This could be valuable information for marketers. Google's new eye tracking technology could be used to reveal your temptations and vulnerabilities, which marketers could then exploit.

* * *

Google Glass enables augmented relativity.[50] Among other claims made about the product, Google Glass is claimed to have the ability to enhance people's lives and performance. For example, Google Glass may make doctors better surgeons.[51] Despite the benefits Google Glass could provide, there are aspects of the technology that Google does not want you to know about.

Google has been very actively pursuing eye-tracking technology.[52] The company was awarded a patent with the innocuous title "Unlocking a screen using eye tracking information."[53] (Hereafter, this title will be shortened to "Unlocking a screen." Note that it refers to unlocking a computer screen after a period of PC inactivity.) Perhaps it should rightfully be titled "Unlocking and extracting information from your mind using eye tracking." Uncannily, drawings accompanying this patent titled "Unlocking a screen with eye tracking information" have many similarities to Google Glass. Also, note the bird (item 402) moving from position A to position B, as this will be discussed in more detail below.

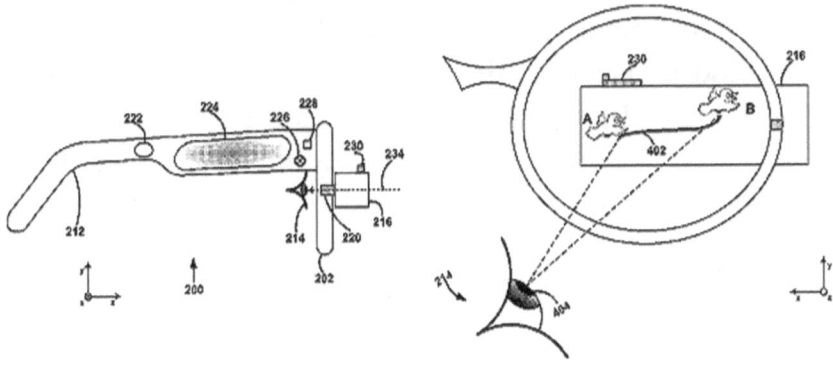

Figure 1 on sheet 1 (of 8) of US patent 8,235,529[54]

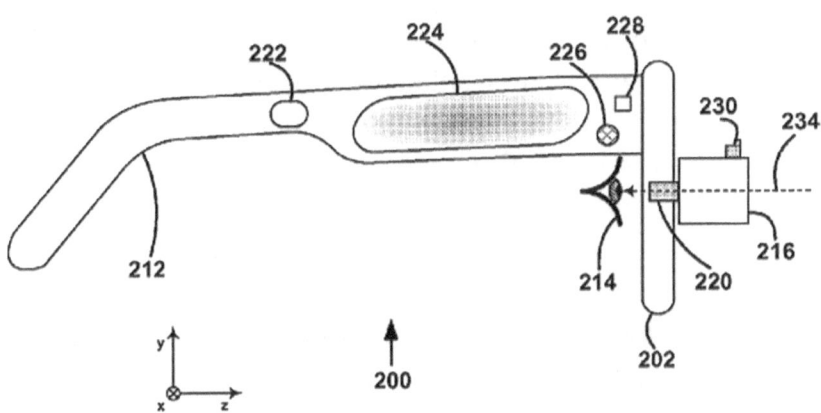

Figure 2B on sheet 2 (of 8) of US patent 8,235,529[55]

Google has applied for a variety of patents to protect Google Glass. Samsung does not have a patent titled "Galaxy S5." Nor does Apple have a patent titled "iPhone." Google is not required to have a patent titled "Google Glass" in order to protect the various technologies that are part of Google Glass, since Google (like other companies) can essentially mix and match patent protection from various other patents it owns.

The Google patent mentioned above ("Unlocking a screen") has many similarities to Google Glass, yet it has a title that most people would not associate with Google Glass. Furthermore, I could find no evidence that Google is marketing any device with the stated purpose of this very peculiar patent, which is for an eyeglass-like device that can unlock a screen. Even though the title is different ("Unlocking a screen" instead of "Google Glass"), is it possible that Google would use this patent to protect some additional features and technology that it may quietly slip into Google Glass or into smartphones? If so, then perhaps we may find some other important clues about the possibilities of reading your mind that Google has, so far, failed to publicize.[56]

Competition in this industry is fierce, so companies are constantly trying to come up with new ways to gather information.[57] Is Google seeking ways to collect[58] very personal information about you in order to sell that information in this competitive market? If Google were able to read people's minds and sell the very private and personal information gathered in this way, then it would have an advantage over its competitors. As just one example of such possibilities, one company has already developed an app that the company claims can detect your emotions by using Google Glass and then relaying that information back to retailers.[59]

Intent is difficult to prove. I am not alleging that Google is currently using eye-tracking technology to read people's minds. Instead, this book is meant to provide a technological assessment of how recently patented eye-tracking technology could be used in hypothetical situations. However, you should be aware that Google's patent for "Unlocking a screen" protects technology that could enable computers to read your mind in some hypothetical situations. The technology could do this by closely monitoring what you're gazing upon (i.e., noting your gaze axis) and the jerky movements of your eyes, the latter of which reveals what your subconscious is watching.

Analyzing a person's covert attention requires a device that can precisely measure the movements of saccades. The text of Google's patent "Unlocking a screen"[60] indicates what is happening when one is in the process of using the eye-tracking device. The process used to unlock the screen is merely a calibration phase that bypasses the cumbersome manual process previously required to use other eye-tracking devices. A manual calibration process provides a warning to users that the device is about to start reading their minds. Google's patented technology, however, automates the previously cumbersome manual process and makes this step invisible to the user.

This patent for unlocking a screen uses the term *eye tracking* eighty-eight times and some form of the word *calibrate* thirty-three times. Of course, there is nothing nefarious about eye-tracking devices in and of themselves. These devices have been around for years, and they have been used for many different purposes, as mentioned above.

Before eye-tracking devices can be used, they must first be calibrated[61] so that the computer can accurately track what the person is looking at. This can be accomplished by instructing a person to look directly at a spot on a screen and then at another spot in a different location, followed by more spots in more locations until the calibration is complete. Alternatively, an individual may be instructed to look sequentially at various letters on a wall chart.

There are two drawbacks to this manual method: (1) it takes time, and (2) it is a sure tip-off to users that their eyes are being calibrated. The latter is not a problem if the person agrees to this calibration. In fact, when using a manual method, it is almost impossible to calibrate a person's eyes without his or her consent and cooperation. However, the methodology described in Google's patent for its device to unlock a screen allows an eye-tracking device to be calibrated without the user's knowledge.

The glasses unlock the screen once a person's eyes are calibrated either by having him or her watch a bird move across the display from point A to point B, as shown in item 402 in the drawing above and in figure 3 of the patent (below), or by having him or her read text, as shown in figures 5 and 6 of the patent (below).

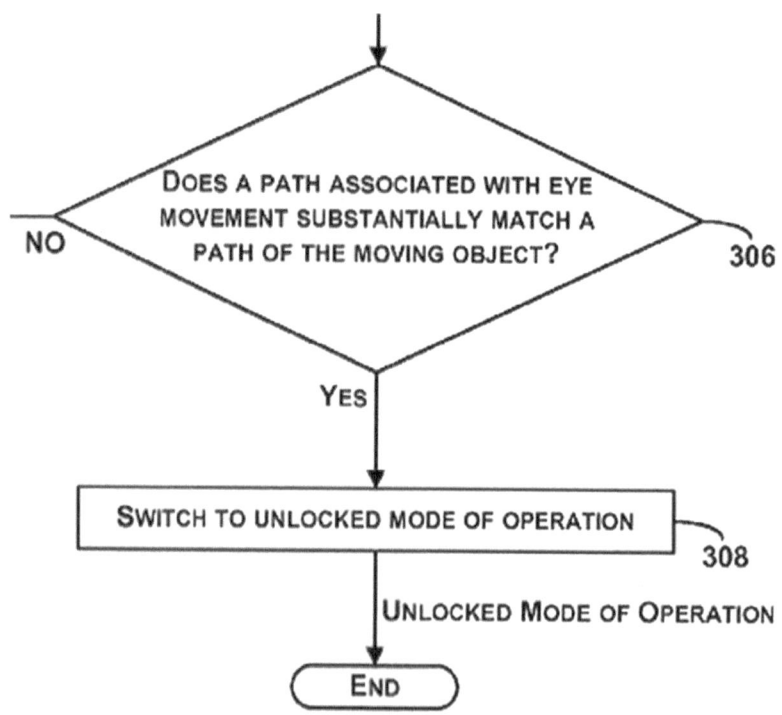

Extracted from figure 3, sheet 3 (of 8), of US patent 8,235,529[62]

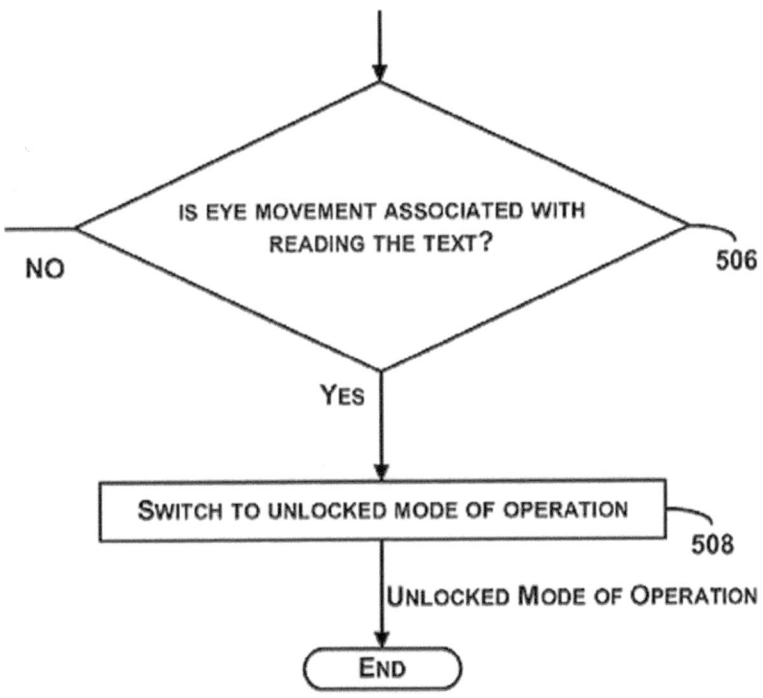

Extracted from figure 5, sheet 5 (of 8), of US patent 8,235,529[63]

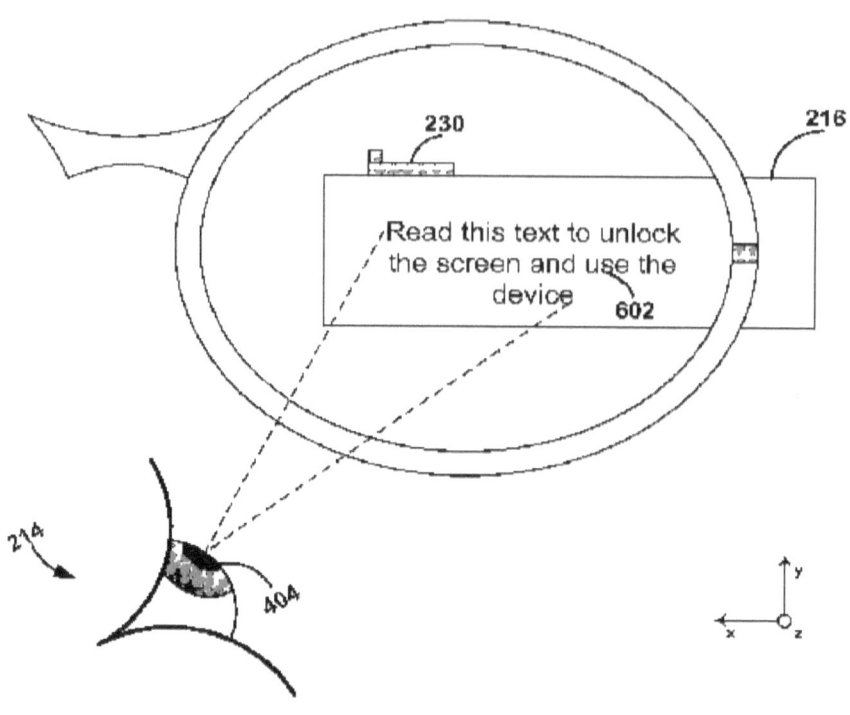

Extracted from figure 6, sheet 6 (of 8), of US patent 8,235,529[61]

Allow me to provide you with some perspective. Google has patented a process to calibrate your eyes every time you read text displayed on a screen. You may think you are only asking for information and then seeing the information quickly pop up on the display, but the device can use the opportunity to calibrate your eye. Therefore, this process allows Google to bypass the manual calibration process and to do so without your awareness or consent.[65] It is not just Google Glass that may be able to use eye tracking to look into your subconscious. Google may use this technology, if the company so desires, in its Android operating system, which is installed in approximately one-third of smartphones.[66]

Eye-tracking technology is a very popular smartphone feature.[67] Check your smartphone to see if it offers this technology. You may be surprised. Samsung Electronics has a smartphone market share between 20 percent and 30 percent.[68] Samsung smartphones and several other smartphones use Google's Android operating system. Samsung Electronics recently introduced EYECAN+,[69] which the company describes as a "second-generation eye mouse." Allegedly, it is superior to Google Glass. "EYECAN+ does not require users to wear any device, such as glasses like Google Glass to calibrate the user's eye."[70] Apple is not sitting on the sidelines. Apple was awarded a patent in January 2015 for eye tracking, tracking a point of gaze, monitoring pupils, and blink tracking.[71]

* * *

The claims made in patents provide the inventors with legal protection, "as it is the claims that define the scope of the protection afforded by the patent and which questions of infringement are judged by the courts."[72] If Google's intention for developing "Unlocking a screen" was merely to provide people with a more convenient way of unlocking their computing systems after a period of inactivity,[73] then the patent could have stopped at claim 8, as that is where the patent describes the screen as being unlocked.

This patent consists of twenty-seven claims, but fewer than one-third of the claims are devoted to explaining how the glasses get to the point of unlocking the screen. So why the extra claims describing other eye-tracking functions *after* the apparatus has unlocked the screen? After all, the title of the patent is "Unlocking a screen using eye tracking information."

Perhaps we may find a clue about what else these glasses can do by looking at the wording in the claims following the description of unlocking the screen. At the end of claim 8,

Google states that after switching the computer from locked to unlocked mode, the glasses can track a user's eye movements. This is not my interpretation; instead, it is Google's own words. The patent language mentions eye tracking many times. It also describes methods to measure the user's rapid eye movements (saccades). This patent contains many more claims than just the one about unlocking a screen after a period of inactivity. This raises many questions.

* * *

With a few modifications, these technologies would not be limited to Google Glass or smartphones. This technology could be scaled up and used in any public place that has a video display. Remember that computers are becoming progressively smaller and that sensors are progressively becoming better. Therefore, any video display in any mall, train station, airport, or other public place could be utilized to read people's minds soon (if this isn't being done already). Consider that any moving objects, such as birds moving across a large public display or text moving across a display, may be used to calibrate the human eye. Couple this with facial recognition technology, and people's minds could be read in any public place that has a video display.

Instead of large public displays, consider the possibilities of miniaturized displays. An accelerometer and gyroscope are inexplicably part of Google's device that unlocks a screen by using information derived from eye tracking. An accelerometer and a gyroscope would be useful in measuring saccades if Google made Google Glass small enough to insert into contact lenses. A miniaturized camera that could move with your eye would be able to record the object that has your covert attention. Although Google Glass has a camera, the camera is focused on those things that have a user's overt attention. It may miss objects in a user's peripheral vision. A camera that moves with the eye would be much better suited for recording those things to which a user pays covert attention if the objects were near or outside the "peripheral vision" of the camera that is part of Google Glass. Apparently, this concept has not escaped Google's attention, as the Internet giant recently announced that its secretive Google X lab has developed a miniaturized Google Glass (including the camera) that is able to fit on a contact lens.[74]

Electronic Mind Readers?

Neuromarketing is used very effectively to influence consumer choices.[1] As mentioned before, many neuroscientists believe that 95 percent of human thinking, including emotions, occurs below the level of conscious awareness, i.e., in the subconscious.[2] Neuroscientists have developed techniques to manipulate the subconscious awareness. One of the goals of neuromarketing is to refine techniques to manipulate the brain's subconscious in order to embed the desired perception as deeply as possible.[3]

Many companies that commercialize technology that can monitor the brain's activity are very squeamish about the phrases *mind reading* and *mind reader*. Instead, they use other terms that downplay mind reading. After you review the recent advances in technology, you should be able to decide for yourself whether the terms *mind reader* and *mind reading* are appropriate. Let us begin by looking at one technology that uses brain waves to analyze what is happening in the brain and where.

Neuroscience has advanced to the point where researchers can very accurately measure activity in specific parts of the brain. This has brought a lot of insight into the thought process, both at the conscious level and the subconscious level. In some situations, these measurements are so accurate that scientists can predict how a person will respond even before he or she says something or moves his or her body. This information is gathered by monitoring signals from the brain. That should not be too surprising. Because the mind controls the body and the mind has to make a decision before the body can execute that decision, scientists, by closely monitoring the inner workings of the mind, can, many times, accurately predict how the body will respond.

An EEG is electronic device that measures various types of brain waves.[4] Because the acronym EEG is too nondescript and the twenty-one-letter word *electroencephalograph* is long, some people refer to this device as a "neuroheadset."[5] The EEG uses antennas to pick up electromagnetic radiation, much like a radio antenna picks up broadcast signals emitted from a radio tower. However, EEGs are able to pick up much weaker signals than those from radio stations. Additionally, EEGs often have special electronic circuits to filter out interference. Brain wave measuring devices are not just for research. Inexpensive EEGs are used to control some games,[6] toys,[7] and mind-controlled music.[8]

You should know that there is a specialized form of EEG called an SST (steady state topography),[9] which is very good at analyzing brain activity. Within the neuromarketing arena, SST is considered the "final word on the human brain"[10] since the SST allows direct measurement of emotions.[11] According to one company's website, the SST "quantifies subconscious and emotional responses."[12] Not only is SST more precise than functional magnetic resonance imaging (fMRI), but it is also cheaper and quicker than fMRI.[13]

CAN A "MECHANISM" ACCURATELY REFLECT YOUR THOUGHTS?

To most people, a device that is able to read human minds seems like science fiction. Scientists conducting basic research into mind-reading technology use terms such as "reading the mind."[1] However, marketers are more squeamish about using such terms, probably because they fear public backlash. Indeed, many people are uncomfortable with terms like "reading the mind" and "mind reader." Perhaps some would feel better using a device that could merely "accurately reflect a subject's actual thoughts."[2]

Does such a device sound like science fiction? It is not science fiction. The US Patent and Trademark Office (USPTO) recently awarded a patent[3] for just such a device to Anantha Pradeep PhD. The patent was assigned to the Nielsen Company. According to Nielsen's website, "Nielsen is a world leader in consumer measurement."[4] That statement seems to be a bold one. However, this device may make that statement very true in ways that few would fully understand.

Words have meanings, and the USPTO is very strict when it comes to the use of words. It has to be. There are many lawsuits filed because of actual or perceived patent infringement. When lawsuits are brought to court, judges and juries look very closely at the words of the patent. Therefore, the words must be precise. What is stated must be able to be proven and verified. There is no hyperbole allowed in the language of patents.

When many people see the word *mechanism*, they think of a machine with gears, pulleys, pushrods, turning shafts, etc. Few would associate the word *mechanism* with something that consists of a collection of advanced sensors and some of the most sophisticated algorithms developed. Few would believe that a mere mechanism could read people's thoughts. Yet the patent for Dr. Pradeep's "mechanism" uses some form of the word *emotion* thirty-eight times, *neuroscience* eight times, *thought* eight times, *feeling* four times, *neuromarketing* three times, and *mental state* three times.[5]

It appears that the attorneys who filed this patent intentionally avoided using the phrases "electronic device" and "mind reader" in order to prevent public concern about the real capabilities of this mechanism. After all, headlines such as "Mind Reader Recently Patented" would probably generate many questions from the public that the patent attorneys' clients would not want to answer.

Many people are aware that the lie detector test (also known as the polygraph test) was invented almost a century ago. The lie detector records a person's blood pressure, pulse rate, breathing rate, and skin conductivity by way of sensors placed on the body that send signals that are translated onto a recording chart. Analog polygraph machines print out this information on a continuously running piece of paper. Although lie detectors provide a lot of information, they are not infallible; therefore, their results are not admissible as evidence in many courts. This is likely what lead DARPA to fund research after 9/11 to design new lie detectors that are more effective.[6] Along with improvements in lie detectors based on better ways to monitor the brain's response, recent discoveries in neuroscience revealed much about where the brain was processing different functions. If someone could somehow link these new methods for monitoring the brain together, they would have a way to gain very powerful insights into what a person was thinking. The problem was that there are different lag times from which the sensors can detect what the brain is doing. To understand this better, let us look at a part of the brain that is called the brain's reward circuit.

Again, I see a problem with the nomenclature. Just as the acronym *fMRI* does not help many people understand why the fMRI device is important to the field of neuroscience, the anatomical term *nucleus accumbens,* being new to most readers, does not help many people understand what this part of the brain is. Unfortunately, neither the English translation nor the Latin root word provides help for most readers who are trying to understand the importance of this part of the brain.

The nucleus accumbens is the portion of the brain that deals with motivation, pleasure, and reward. It also plays a central role in addiction.[7] This area of the brain is known to marketers and neuroscientists as "the craving pot."[8] If you want to know more, then be aware that Brian Knutson, PhD, of Stanford University has a video[9] on YouTube that shows the location of the nucleus accumbens and describes some of its function. If marketers can refine an advertising campaign that causes this part of the brain to light up (when viewed with fMRI) in volunteers, then many consumers will find the product hard to resist. Recall that Martin Lindstrom described the fMRI as "a killer marketing tool."[10]

* * *

The facial action coding system (FACS) was mentioned before in this book. FACS is described as "a procedure to systematically describe human facial expressions."[11] What is more important is that computer algorithms have already been devised to read facial emotions.[12]

Anantha Pradeep, PhD, was awarded a patent in 2013 titled "Audience response analysis using simultaneous electroencephalography (EEG) and functional magnetic resonance imaging (fMRI)." In this process, brain waves, "facial emotion encoding," eye-tracking data, and physiological data can be "analyzed to provide measures of *emotional engagement*" (author's emphasis).[13] Additionally, people's facial expressions, automatically recorded by computers, can be assessed to increase the accuracy of measuring people's emotional engagement with what they're seeing.[14]

Most people cannot accurately determine another human being's actual thoughts in every situation. As Dr. Pradeep explains in his patent, many conventional automated devices attempted to detect thoughts, but had problems with accuracy.[15] Ironically, this is very much like a common human problem—just ask any couple who have been married more than a year.

Nevertheless, as Dr. Pradeep explains in his recent patent ("Analysis of marketing and entertainment effectiveness"), a significant problem with conventional systems is that these "mechanisms are severely limited in their ability to accurately reflect a subject's actual thoughts."[16] However, it appears that Dr. Pradeep and his group have found a way for computers to overcome the deficiencies of previous mechanisms. It is important to note here that the phrase "accurately reflect a subject's actual thoughts" is used in this particular patent that was granted by the USPTO. The reason why the USPTO awarded a patent to Dr. Pradeep and the other two inventors is that they were able to significantly improve the ability of a mechanism to "accurately reflect a subject's actual thoughts." Again, please note that the USPTO is very strict about the use of words and does not allow hyperbole.

Dr. Pradeep's patented device cleverly sequences the information coming in from different sensors and integrates them properly in time. Let's review some of the different sensors and learn how they gather information at different times according to how the brain processes information.

There are five different types of brain waves, and these each provide different types of information. The brain waves are categorized by the length of time it takes for the brain wave to start and finish. Alpha brain waves are slower (eight to twelve transitions, or cycles, per second[17]) than gamma brain waves (forty to seventy transitions, or cycles, per second[18]).

Alpha brain wave activity often precedes gamma brain wave activity, and alpha brain waves can precede a final decision sometimes by several seconds. Alpha waves are often associated with enhanced problem solving.[19] Gamma brain waves are "able to link and process information from all parts of the brain."[20] When the mind is thinking, alpha waves are abundant. When the mind has solved a problem at the subconscious level, a burst of gamma waves indicates that the solution or decision has been moved to the conscious awareness.[21] Think of the slow alpha waves as long, drawn-out deliberations and of the fast, jumpy gamma waves as giving rise to, "Aha, I finally have the answer! I have to tell everyone as soon as I can."

As brain cells become more active, they need more oxygen, so blood flow is increased to provide the cells with oxygen. Increased blood flow as measured by fMRI often lags behind increased brain activity by about six seconds.[22]

The mechanism developed by Dr. Pradeep and his group is able to properly sequence results that come in at different times from EEG, fMRI, various eye-tracking sensors, video analyzers of facial emotion encoding, and various other sensors. All of these results are synchronized in time (i.e., they are time-phased) to align with what the person saw, heard, smelled, and even tasted and touched. The results provide amazing insight into what the brain was thinking at the time.

After five years of studying the patent application, the USPTO agreed with Dr. Pradeep that his mechanism was a significant improvement over other techniques and methods that attempt to "accurately reflect a subject's actual thoughts." On July 23, 2013, the USPTO awarded a patent to Dr. Pradeep and two other inventors, assigning the patent to the Nielsen Company.

This recently patented mechanism has the potential to turn computers into mind readers. Using the processes outlined in the patent, a computer can take information from multiple sensors and modalities[23] and synthesize it all, which greatly increases the computer's ability to accurately reflect a person's actual thoughts. Recall that Google's CEO Eric Schmidt stated, "We [Google] can more or less know what you're thinking about."[24] This comment from 2010 was based on inferences that Google is able to make based on users' Internet activity. Couple those accurate inferences with sensors on a smartphone that can monitor in real time a person's eyes, facial expressions, and voice, and you can see how the smartphone can also be a very accurate mind reader. Eric Schmidt clarified Google's aggressive policy by saying, "Google policy is to get right up to the creepy line and not cross it."[25] You can make your own decision about whether or not Google has crossed or will soon cross the creepy line.

To put this into perspective, there is now a system of sensors and software that can do what Frankenstein wanted to do but was unable to do while quietly remaining in his hovel. Have the tables been turned on us? Perhaps what was science fiction has now become reality. Instead of reading a science fiction novel, perhaps someone–or something–is reading you like a book as if you were one of the lovely creatures that Frankenstein observed from his hovel. If you are reading this[26] on your smartphone or a tablet, contemplate what it would be like if a Frankenstein-like system had been monitoring your eyes, facial expressions, and voice as you read up to this point. Has your smartphone, laptop, or PC been acting funny or running slow lately? Do you have a feeling you are being watched?

* * *

Of course, the Nielsen Company is not able to use Dr. Pradeep's device mentioned above to monitor a person without first obtaining his or her permission or consent. Additionally, this device is likely too big and bulky to move easily. However, Nielsen could do the next best thing. They could use the device on large groups of people—audiences—who are presumably volunteers and test their emotional responses to "a media clip, a commercial, a brand image, a magazine advertisement, a movie, an audio presentation, particular tastes, smells, textures, and/or sounds."[27]

In doing so, the device could determine the emotional responses of the various volunteers and sort these by the volunteers' age, gender, ethnicity, income level, political orientation, sexual orientation, religious orientation, education level, etc. Then, Nielsen would be able to refine advertisements, brand images, audio presentations, etc., so they would have a very high likelihood of eliciting strong emotions in people or activating the brain's reward circuit like cocaine does for an addict. In this manner, marketers can use this mind reader to refine advertisements, brand images, audio presentations, etc., by gauging their effect on volunteers who are very similar to you. Once the advertisements, etc., were perfected, you could be targeted.

Several data brokers offer a service called "onboarding." Onboarding typically includes three steps: (1) segmentation, (2) matching, and (3) targeting. The onboarding process starts with segmentation, when a client asks a data broker to find consumers who exhibit particular characteristics.[28] Characteristics may include such things as ethnicity, income level, education level, sexual orientation, political association, etc.

A Federal Trade Commission (FTC) document reads, "The next step in the onboarding process is 'matching,' where the data broker finds the consumers it identified through the segmentation process online. To find consumers online, the data broker enters into contracts with registration websites to buy lists of registered users. It then compares these registered users with the consumers identified through the segmentation process in order to find matches."[29]

The FTC document goes on to state the following:

> The last step in the onboarding process is to target the matched consumers online. To do so, the data broker must first place an HTTP cookie[30] on the browsers of the consumers it has identified through the above process. It does so when the registration website notifies the data broker that such a consumer has logged on to the registration website. The cookie includes the information that the data broker has appended to the consumer's profile.[31]

Marketers can target people with advertisements, brand images, audio presentations, etc., that are cleverly designed to slip past a person's rational conscious mind and move into his or her subconscious mind. In this manner, marketers achieve a very high likelihood of eliciting a strong emotional response in potential customers. Whereas one may think that he or she is merely looking at an interesting picture or listening to an interesting audio presentation, one is actually being specifically targeted with a well-crafted way to elicit emotion. After seeing or hearing the advertisement (or other appeal), many might assume that they just discovered, by chance, something cute or interesting. On the contrary, the targeted person is deliberately being drawn into a psychological ambush designed to elicit a response by using thoroughly tested appeals aimed at the subconscious. Moreover, people often have less money after experiencing these psychological marketing ambushes (meaning that people spend their money to purchase the product or service being advertised).

If you are active on social media or if you blog frequently, then marketers may target you, with the expectation that you will "tout these products to [your] friends and followers." [32] In this manner, marketers use you to persuade your friends to buy certain things. Marketers may not have read your mind, but they have carefully read the mind of someone who has been identified as a very close match to you. You may also be targeted with the expectation (as mentioned above) that you will "tout these products to [your] friends and followers."[33] Apparently, this tactic works often enough that advertisers are willing to pay data brokers for lists of those people who are active on social media and blogging sites.

"Wearables": For Your Health or for Their Profit?

Wearable health-monitoring smart devices are the next important market that many are trying to tap.[1] An estimated twenty-one million wearables were sold in 2014.[2] The 2015 sales estimates for the Apple Watch range from ten million to thirty-two million units worldwide.[3] Many corporations are scrambling to find ways to provide people with more information about their health. Smartwatches with heart rate monitors will be able to track a person's daily activity. They will also be able to monitor a person's sleep patterns and determine whether he or she is resting well or tossing and turning all night.

Health monitoring devices can monitor a lot about a person, including a wide range of motions, physiological data,[4] and location data. Health monitoring devices can then sell this information to data gathers who, in turn, sell it to data analyzers who, in turn, sell it to marketers. Several possible sensors might be incorporated into smartwatches. These include accelerometers, gyroscopes, magnetometers, barometric pressure sensors, an ambient temperature sensor, a heart rate monitor, an oximetry sensor, a skin conductance sensor, a skin temperature sensor, and a GPS.[5] The Apple Watch includes an accelerometer, a gyroscope, a heart rate sensor, and a barometer.[6] The accelerometer and gyroscope can work together to function like a pedometer and count a person's steps. However, cycling is smoother, so Apple Watches rely on the GPS and Wi-Fi in iPhones to measure the distance a person travels by bicycle.[7]

Smartphones have evolved to the point where they can scoop out as much information as possible by using the electronic platform. This is why many corporations are now seeking to gain more information about people by monitoring their health.

While it may seem to be advantageous for people to have an increased ability to monitor their health, there are some potential downsides to wearables that provide health-related data. The issue of privacy is starting to concern many.[8] For example, the new iWatch may allow employers to snoop on employees.[9]

If the previous trends in advertising hold true, then we can expect some marketers of health-monitoring wearables to use fear as a marketing tactic. Some people will be able to improve their athletic performance by closely monitoring their physiological processes with these devices. Although people may get into better shape with this extra monitoring, it is possible that they will soon be much more apprehensive about their overall health. Some marketers may flood these people with advertisements claiming that the devices can prevent an array of diseases and ailments. Recalling the earlier chapter about fear, fear is a powerful motivator and is not accurately correlated to real probabilities. Therefore, the outcome of people's use of smartwatches that monitor health may be people who train harder and are in better shape, but who become much more apprehensive or fearful about their health.

* * *

In the early 1970s, when I was still in high school, I began programming computers by using the Fortran programming language. One of my high school teachers, Mr. Pearson, was a visionary. He went out of his way to start a computer programming course for the students at his school. He had to obtain permission from the school board and to get permission from the local community college to use its large IBM mainframe computer to run student-created programs. We students used IBM punch cards, which we punched one at time with a leased IBM card puncher that was larger than an upright piano. Mr. Pearson arranged a tour for us to go to the community college and see the IBM punch card readers, the large mainframe computer, the magnetic tape drives, etc., all of which filled a very large room.

Mr. Pearson saw the future. He told his students that we should be on the lookout to buy stock shares offered by start-up companies that promised to bring computers to consumers instead of just large businesses. Back in the 1970s, Mr. Pearson told us students that everything we saw in that big room would someday be made small enough to fit into a bread box. One of the students laughed and said, "And I suppose people will be wearing Dick Tracy watches too!"

Mr. Pearson was right on the money, and he knew it. He simply replied, "Mark my words."

The class clown did not realize how correct the teacher was, either. It would take the former almost four decades to realize the truth in Mr. Pearson's words, even though Mr. Pearson had made his statement offhandedly. I was that class clown. By the way, I also ignored what was probably the best stock tip of my life (that of small computers becoming consumer items). I suspect Mr. Pearson bought himself a very nice mansion on a beach somewhere. Here is a call-out to Mr. Pearson: I should have paid closer attention to what you said!

Someday Soon, Maybe Very Soon

Chapter 8 of Bruce Schneier's 2015 book *Data and Goliath* illustrates how "the manipulation [done by corporations] is largely hidden and unregulated, and will become more effective as technology improves."[1] How could these new technologies play out in the lives of consumers? The technology in *Minority Report* existed in 2054, but we may not need to wait that long. Let us explore some possibilities based on currently existing technologies and on some technology that may be developed soon. For legal reasons, I am not alleging that corporations have been or are currently engaged in activities similar to those described below. Instead, I seek to provide a technological assessment of corporations' capability to do the things described below. You may make your own inferences about whether or not corporations would pass up the huge profits made possible by recent technological developments.

This part of the book received a wide variety of comments from reviewers before it was published. Some wanted to see more scenarios illustrating how these technologies could affect consumers, while other reviewers wanted to see fewer scenarios. You may want to read some or all of the following scenarios before starting the next chapter, "Legal and Ethical Issues."

Oh Là Là Latest Ladies' Fashions[2] is a hypothetical start-up company that caters to women of all ages. Oh Là Là Latest Ladies' Fashions (hereafter shortened to Oh Là Là Fashions) has recently achieved huge increases in market share in the women's clothing market. The company's stock prices are rising dramatically. How did Oh Là Là Fashions do it?

They did it by carefully integrating what women thought about themselves into online advertisements. They also decreased the cost of the clothing by deliberately choosing manufacturers that would provide adequate-quality clothing without the high degree of fit and finish found in higher-quality brands. Oh Là Là Fashions also chose manufactures that provided clothes to midprice national retailers, but Oh Là Là Fashions sold these clothes to customers at prices just slightly lower than the cost of designer clothing.

A woman named Jane made a comment on a social media website several years ago to one of her friends: "I am not as well-endowed as you are." By analyzing this comment posted on a social media website long ago, algorithms accurately inferred that Jane felt that her bust was not large enough. Recently, Jane has been going to a health club so that she can develop a more shapely body. Jane has never made a comment on social media about going to the health club or about her desire to lose some weight and tone up her body. However, because of the information relayed from her smartwatch, algorithms have figured out that she is primarily working on her legs. This information was inferred by her body movements as recorded by accelerometers and the gyroscope in her smartwatch. These can tell the whether Jane is working on the StairMaster or is participating in a yoga class. Although Jane is interested in yoga, her primary interest of late has been to tone up the muscles in her legs so her legs become shapely.

In order to "provide a more enjoyable online experience," Oh Là Là Fashions' software, using techniques developed by filmmakers to replace missing actors (e.g., computer-generated images, or CGI),[3] automatically transforms two-dimensional pictures to create a three-dimensional model[4] of Jane. When Jane shops online, she uploads several photographs showing her wearing her yoga outfit. (Jane had to pose in her yoga outfit as one of her friends used a smartphone to take forty or more pictures of her from different perspectives and from three different heights.)

All Jane has to do is enter her height and upload the photographs. The computers and algorithms autonomously do the rest. Algorithms adjust Jane's 3D avatar based on inferences made about her desired self-image that are gleaned from her comments posted on social media websites, her online browsing history, her eye gaze activity, and her exercise activity. The team of crack marketing psychologists and programmers have developed algorithms that adjust Jane's 3D image so it reflects how Jane would like her body to look. Algorithms autonomously resculpt her body so it appears twenty pounds lighter. In addition, her bust is larger and her legs are shapelier. Additionally, her facial expressions are reworked using facial emotion algorithms so that the image of her face on the shopping website exudes happiness and self-confidence.

Jane can electronically try on clothes. When she does this, she finds that the clothes fit perfectly and are very flattering. However, she notices that, when it comes time to order an article of clothing, she does not need to select a size. The algorithms correctly order the right size for her, even though the online videos and still pictures show her wearing clothes that are smaller and that nicely fit her electronically resculpted body. The algorithms cleverly show her what she wants to see, but the company sends her the clothes that will fit her body comfortably. In this manner, Jane feels less angst about entering her real size when she orders clothes. When the new clothes arrive, she tries them on and likes how they fit—and she was able to bypass the awkward stage of entering her real size.

Oh Là Là Fashions has made an agreement with its clothing supplier to print on its clothing labels a size that is one size smaller than the actual size of the clothes. This makes the customer feel better about herself when she wears the clothes. More important, however, is that this subtle trick greatly improves customer loyalty and ensures repeat business. When customers go to brick-and-mortar stores to try on clothes, they find that the clothes whose labels show the same size as those from Oh Là Là Fashions feel tighter. The customers' shopping experience is unpleasant when they are confronted with reality, so they prefer shopping with Oh Là Là Fashions, even though some customers may suspect that the sizes printed on the labels are incorrect.

Advanced algorithms have prepared a gallery of pictures and videos of Jane. In these still photos and videos, Jane is shown in her normal surroundings: the place where she shops for groceries, her place of employment, her gym, and her favorite restaurant, disco, nightclub, and so forth. Her face, exuding confidence and happiness, is made possible by the facial emotion coding algorithms. The videos even show Jane walking through her favorite restaurant, with many admiring people turning their heads to look at her. Her beauty, her shapely, resculpted body, and her electronically fitted clothes capture the attention of many people as she strolls through the restaurant. Some of the admiring people sitting at the tables are her friends. Their faces have been identified on social media. The photographs of their faces are transformed by algorithms so they reflect an admiration of Jane. Jane likes what she sees, so she buys the clothes. Moreover, she likes what she sees so well that she buys more clothes than she used to buy when she went to the mall to shop.

Oh Là Là Fashions' corporate strategy is to buy clothes at a midrange price and then sell them at higher prices than other retailers charge by pitching a very strong appeal to the consumer's subconscious. To do this, the company spends a lot of money to buy a lot of very detailed personal information, including mood data, from data brokers. Oh Là Là Fashions also made a corporate decision to hire a large and talented team of crack marketing psychologists to use this information in the most effective ways. The crack marketing psychologists designed several different advertising campaigns and then tested them out on volunteers.

These paid volunteers were recruited from different segments of the female shopping population. They were offered a $50 voucher in exchange for evaluating several proposed advertising themes, slogans, pictures, videos, and website features. They were monitored with fMRI, brain wave detectors, facial emotion detection, eye tracking, etc., as they reviewed the advertising promotions. As a result, the team of marketing psychologists were able to refine the advertisements in ways that made them hard for women to resist. It is not as if someone were holding a gun to the women's heads to make them buy the clothing. Nevertheless, the marketing psychologists knew that if they could elicit a strong emotional response or stimulate the nucleus accumbens, then many women would buy their products.

Oh Là Là Fashions does not have a single slogan. They have several, depending on which type of woman logs in on their website. Based on the information that Oh Là Là Fashions has purchased from data brokers, the algorithms correctly identified that although Jane has strong concerns about her bust size and her legs, she is not narcissistic. Therefore, the online shopping site sets a tone based on her psychological profile. Jane notices the slogan prepared for her: "We Can Dress You and Make You Look Beautiful."

During the tests involving the paid volunteers, another slogan, "We Can Dress You, Give You Plenty of Hugs and Kisses, and Make You Look and Feel Beautiful," was more popular. However, the marketing psychologists realized that the Oh Là Là Fashions website could not give plenty of hugs and kisses because it was, after all, just a website. One clever marketing psychologist suggested adding pictures of people who looked very similar to an online shopper's mother and father next to the slogan.

The pictures shown during the testing phase were modifications of actual pictures of volunteers' mothers and fathers that were changed just enough so that the paid volunteers could not tell if the photos were of their actual parents. However, the images were similar enough to elicit an emotional response. More ingenious, however, is that the facial emotions algorithms transformed the faces so the people pictured radiated interest and joy. This led the volunteers to remember fondly the times when they were children, when their parents were happy to see them and had time to play with them.

The combination of the shorter slogan and the images of approving and loving parents scored almost as high as the longer slogan by itself when the psychologists monitored the volunteers' nucleus accumbens and their brain's emotional response. Although many of the paid volunteers expressed feelings ranging from disappointment to bitterness toward their often-absent fathers and tired, overworked mothers, the images of loving and caring parents overcame those negative feelings. This is similar to Leo Burnett's clever use of pitches to the subconscious to overcome men's fierce resistance to the Marlboro brand.

The slogan customized for Jane appeals to Jane because her mother was often preoccupied with work and with Jane's siblings, including one who needed a lot of extra attention because of a disability. Jane's father was often absent. Jane's mother was kind, but she was just too busy to give much time or attention to Jane. The slogan pitched to Jane hints of caring parents who will take the time to dress the young Jane properly before she goes out to face the world. It appeals to Jane's subconscious belief that she was an unwanted child. Now she has found something that will provide her the care and nurturing she craved as a child.

Jane also likes some of the other features on Oh Là Là Fashions' website. In addition to her cleverly resculpted body, several other features allow Jane to see how the clothes would look on her if she had various lengths and colors of hair as well as different makeup effects. The initial 3D avatar of Jane is based on a series of recent pictures taken with her smartphone and at different angles. However, her makeup is adjusted by algorithms to make it appear as if she had been made up by the finest makeup artists in Hollywood. There are slider bars that allow Jane to change the effects.

Jane can see how she would look with more or less of a suntan, longer or shorter hair, straight or curly hair, and various hair colors. Jane likes to see how she could appear to others, given the right touches. She also likes exploring with various makeup themes and hair lengths without going through the trouble of cutting her hair and then waiting for it to grow out.

Now Jane buys almost all her clothes from Oh Là Là Fashions because the company has successfully activated her nucleus accumbens by showing her images that stimulate positive emotions. Additionally, the team of marketing psychologists developed algorithms so that Jane, when shopping on the website, would feel as if she were getting the attention she has craved since she was a child. Because her psychological desires were identified by marketing algorithms, Jane has a more pleasurable "shopping experience" online than she does at a chic clothing store in the nearby upscale mall.

Jane could save 40 percent on her clothes if she bought from other retailers whose clothing is supplied by the same manufacture that supplies Oh Là Là Fashions. However, Jane simply feels better wearing the clothes purchased from Oh Là Là Fashions. She has a better image of herself because of her resculpted body and face, cleverly provided by using algorithms that determined what her subconscious mind wanted her self-image to be.

Jane also feels more connected to her tired and busy mother and her often-absent father when she wears clothes purchased from Oh Là Là Fashions. When asked, Jane cannot explain why she feels more connected to her mother and father when she wears the clothes, but the team of marketing psychologists know exactly why. Over three weeks, they offered many different slogans to the paid volunteers. After sixteen attempts with different slogans tested on 112 volunteers, the crack marketing psychologists finally found a slogan that resonated well with customers like Jane. That may seem like a lot of time and money just to decide on a slogan. However, the payoff is huge when marketers get it right and create a pleasurable shopping experience.

* * *

Peter is a middle-aged male that is successful in his profession. Information gathered from his smartphone indicates that he has not been getting the romantic attention from his wife that he feels he deserves. As a result, Peter can expect to receive advertisements for escort services, pornography, and alcohol or marijuana (in states where the latter is legal).

When Peter visits pornography sites, his activity and mood are monitored closely. Other pornography purveyors will be willing to pay a premium for information about where his eyes look and for how long. If Peter appears to be getting tired of soft pornography, then he can expect to receive offers from purveyors of hard pornography. If he becomes addicted to pornography and wants to escape its clutches, then this will be difficult for him, because the data brokers have identified him as a potential pornography addict; therefore, he will continue to be bombarded with ads for pornography. It will be unpleasant for him to use the Internet in the future because of all the advertisements he will see for pornography.

One day, Peter gets an advertisement from a bail bondsman and thinks it is crazy. He has never been arrested. However, that night, he gets drunk, drives his car, gets in an accident, is arrested, and is jailed. After he sobers up and is released from jail, he remembers that the day before he had gotten the advertisement from the bail bondsman. He quickly finds the advertisement and calls the number. Is this a coincidence? No.

Algorithms detected many recent changes in Peter. Eye-tracking information from his smartphone indicated his new interests. His smartphone, using Apple's newly patented process, detected his mood change. His smartphone also detected changes in his mood by analyzing his voice. Facial expression software also detected a significant change in his facial expressions. All of these devices detected his moderate mood shifts when he was sober, and his huge mood shifts when he was drinking.

The smartphone and algorithms correlated Peter's mood shifts with his purchase of alcohol, since he used electronic money and his smartphone to buy alcohol. Data from cell phone towers also pinpointed him at various bars around town and discerned that he was staying at bars longer than he normally did. Additionally, his online viewing of pornography had recently changed. Predictive algorithms identified that he was at higher risk now for overindulging in alcohol—and a bail bondsman was willing to pay for that information. It would have been nice if bail bondsmen had offered to intervene the day before to help Peter address the problems he was experiencing. However, that is not how bondsmen make money.

* * *

Kathy is an attorney. She played varsity volleyball while in college. However, now she has little time to exercise because of the time she commits to her job and to her family. She has gained twenty-five pounds and is now ashamed of herself. All of this information can be quickly put together after analyzing different pieces of information purchased through data brokers, even if Kathy never talked about her problems or sent a text message specifically using words to describe her weight or the shame she feels. The algorithms can take millions of seemingly disconnected pieces of information and make very accurate inferences from them. As a result, Kathy can expect to receive advertisements for weight loss programs, health clubs, and yoga classes, as well as books with titles that promise enhanced self-esteem and the development of a good feeling about herself.

When Kathy goes to shop online at Oh Là Là Fashions, she likes what she sees. The algorithms produce still pictures and videos of her more shapely body wearing the clothes. Her face, as shown in the still pictures, radiates confidence, excitement, and pride, which takes her back to her college days when she played volleyball at the peak of her athletic ability. Now Kathy buys almost all of her clothes from Oh Là Là Fashions' website because it has successfully stimulated her nucleus accumbens. Kathy has a more enjoyable shopping experience online than she does in a store. Additionally, she does not have to fight for a parking spot or worry about her car being scratched.

* * *

Joe is a teenage boy. He has enjoyed a good life so far, one characterized by adequate self-esteem and self-confidence.

However, Joe has recently experienced several unfortunate life events that have taken a significant toll on both his self-esteem and his self-confidence. Algorithms detect a sudden change in Joe and show that he is struggling with these two issues. Joe now receives advertisements that are pitched to his subconscious and that show images indicative of increased self-esteem and self-confidence.

Instead of providing Joe with a broad approach, such as the one Leo Burnett used for the Marlboro campaign, algorithms customize an advertisement campaign just for Joe based on accurate inferences made about him. Algorithms are used to adjust Joe's facial emotion coding so that, in the altered photos he sees of himself, he sees his face radiating self-confidence and self-esteem—with the products promoted by the marketers prominently displayed close to Joe's smiling face.

The products displayed in these ads may be varied, depending on who is willing to pay the most for the information that Joe's self-esteem and self-confidence have recently taken a huge hit. For example, he could be sent a video that shows him moping around until he starts drinking an energy drink—at which point everything quickly changes. There is a new spring in his step. Holding the energy drink in his hand, he merrily interacts with his friends as they all take a break from playing their favorite sport. Joe's face radiates renewed hope and enhanced confidence, vigor, self-esteem, and self-confidence. It is as if Joe put on a Superman outfit and became Superman.

Alternatively, Joe could be presented with images showing him to be smiling and self-confident as he wears clothes from a clothes maker that paid a premium for information about Joe's mood. On the other hand, a happy Joe could be shown using the latest smartphone or wearing a smartwatch, smart glasses, etc. The line of products weaved into the images of a happy Joe whose face radiates self-confidence and self-esteem is probably limitless.

These appeals will not be limited to Joe. They will also be cleverly pitched to males and females of all ages who struggle with self-esteem or self-confidence after experiencing the unfortunate events that life throws their way.

* * *

Barbara, a grandmother, is in her midsixties. She fondly remembers her younger years. However, time and circumstances have taken a toll on her body. She notices that she has more wrinkles and that parts of her body are starting to sag. When Barbara goes to shop online with Oh Là Là Fashions, she likes what she sees.

As was the case for Jane, algorithms adjust Barbara's 3D avatar based on inferences made about her desired self-image that have been gleaned from her comments posted on social media websites, her online browsing history, and her eye gaze activity. The algorithms have adjusted her face and body to make her look much younger and more vibrant. She may not have found a real fountain of youth; however, she has found an electronic fountain of youth that is realistic enough. She likes the sweet little electronic lies. Algorithms produce still pictures and videos of Barbara electronically modeling the clothes. These pictures show her face and body as if they had been reworked by a team of plastic surgeons who have done so many procedures on her that she, instead of paying the surgeons, could have purchased a luxury car. Barbara knows that what she is seeing is a fantasy, but she likes it anyway.

Now Barbara buys almost all of her clothes from this website because it has successfully stimulated her nucleus accumbens. Barbara has a more enjoyable shopping experience online than she does in a store, even though she is paying almost designer clothing prices for clothes that are produced by the same manufacture that sells to national midrange retail stores.

* * *

Joyce tends to be a binge shopper when she is in the mood. She has racked up huge credit card bills and is now deeply in debt. The team of crack psychologists at Oh Là Là Fashions designed algorithms to catch Joyce when she is prone to binge shopping. They did this by gathering information about her moods from her smartphone and then correlating this information with data about her credit card activity. As a result, the team of marketing psychologists realize Joyce's potential for being a big-spending customer. With this in mind, they add several features to the website that will specifically play on her recent moods by offering her messages dealing with self-esteem, acceptance by others, romance, or whatever else affects Joyce's mood when it changes from her baseline mood.

* * *

Romantic Vacations and More[5] is a hypothetical resort destination specializing in romance. This company had hired a team of crack psychologists to help improve its marketing methods. The results were spectacular. The psychologists created algorithms to monitor social media websites and correlate the information gleaned therefrom with mood data for single women who are looking for romance. Using techniques like those implemented by Oh Là Là Fashions, the website of Romantic Vacations and More imports pictures of potential customers and then modifies their bodies according to the women's subconscious desires. Romantic Vacations and More transforms these images in the still photographs to show the potential customers at the resort. Like Oh Là Là Fashions, Romantic Vacations and More has algorithms that can generate videos of various scenes, including strolling on the beach at sunset, riding white horses through the waves along the shore, and dancing to lively music until the early hours of the morning. All of this information regarding the website users' preferences has been purchased from data brokers and data analyzers. The women who visit the website see their faces radiating exhilaration, self-confidence, and happiness. They also notice that their bodies have been electronically resculpted according to their subconscious desires. Although they all realize their new looks are a bit fantastical, they all like what they see.

The price of a vacation at Romantic Vacations and More is about 30 percent higher than the price to stay at comparable destinations. However, many women like what they see in the still photographs and the videos, so they are willing to incur the extra cost for Romantic Vacations and More merely because the team of marketing psychologists have successfully stimulated their nucleus accumbens. This is not a rational decision, but rather a decision that has been carefully elicited by marketers by appealing to the customers' emotions.

* * *

Nancy's world is upside down. She is not sure how she ended up in this dismal state. She is talking to her therapist because she is about to be fired from her job. She has huge credit card bills from her compulsive online shopping—especially for clothes from Oh Là Là Fashions.

Nancy has narcissistic tendencies. The algorithms designed by Oh Là Là Fashions' team of psychologists have accurately inferred Nancy's narcissism. Therefore, Nancy sees a different slogan than Jane sees. When Nancy logs on, she sees a slogan more closely aligned to her personality: "Beautiful Clothes for Beauty Queens and Lovely Ladies Who Are Just Too Humble to Wear a Tiara Every Day." Although Nancy has narcissistic tendencies, she does not want to admit it. Still, she likes the idea of being included with beauty queens. She also likes the second part of the slogan, as she realizes that many people do not like open displays of narcissism.

Nancy made one comment to a friend on a social media website long ago about her "big chin," so algorithms carefully adjust her chin when it appears in photos and videos on the website. Nancy really likes the clothes that she can see herself modeling online. Her conscious mind may not be aware that the algorithms resculpted her chin, but her subconscious is aware of the more beautiful face.

Whereas Jane simply enjoys the feature of the website that allows her to quickly change her hairstyle and hair color as well as see the effects of different makeup, Nancy has found the feature to be irresistible. At first, it was just a cute feature that Nancy enjoyed playing with, but within a year, it became an obsession for her.

Oh Là Là Fashions offers five levels of "rewards" for loyal customers: silver, gold, platinum, diamond, and diamond-plus. Once Nancy had spent $5,000 on clothes purchased from Oh Là Là Fashions, she earned silver-level rewards, which included the ability to add accessories and personalized destinations, such as her favorite restaurant and nightclub, to her still photos and videos. The accessory options allowed Nancy to see what she would look like when wearing certain clothes along with accessories like shoes, hats, tiaras, and jewelry. She was also given the option of adding a beauty queen sash.

Sales of authentic Miss America sashes are restricted. "These exquisite, handmade sashes are orderable only [by] authorized persons."[6] At the gold rewards level, Nancy can see her avatar wearing the sash of the beauty queen for a city such as Los Angeles or Seattle, even though she never won (or even entered) a pageant. Once she spends $7,500 on clothes, she is elevated to platinum level, which provides her with the option of seeing her avatar wearing a beauty queen sash for a state such as California or New York. When she spends $10,000, she is elevated to diamond level and becomes eligible to wear a Miss America sash in the still photographs and videos. However, she cannot wear a Miss Universe sash unless she spends $15,000, which is the requirement for advancing to the diamond-plus level.

Over the past several years, Nancy has spent over $12,000, so now, just by sliding and tapping her finger on the option bars in the display window of her smartphone, she can see how she would look wearing a Miss America sash over various clothes and at her favorite restaurants—or at personalized locations, like Paris.

Nancy likes another "customer loyalty" offer that was not offered to Jane—3D sculptures. (This feature was not offered to Jane because she did not show any indication of narcissism.) At the silver level, Nancy received a twelve-inch-tall sculpture of herself. This beautiful replica of Nancy came complete with an expert makeover on a par with the work of the finest makeup artists in Hollywood. Nancy could choose a sash representing the state for which she wanted to be the beauty queen.

Once Nancy earned platinum-level status, she received a twenty-four-inch-tall sculpture that was made of fire-glazed ceramic and that featured a painted-on tiara. At the diamond level, Nancy received a thirty-six-inch, glass-infused nylon 3D painted sculpture of herself. However, it included a separate, removable tiara so she could place the tiara on her alter ego, as if someone were coronating her after she won a beauty pageant. This tiara features with small, synthetic diamonds that sparkle like real diamonds. The sculpture shows teeth that are pearly white even though Nancy's real teeth are not quite that white. Facial action coding system algorithms ensure that her face lights up like the faces of Miss America winners. The sculpture is beautiful. If Nancy resembled the enhanced sculpture, then she would likely win a real beauty pageant.

Nancy can watch videos of herself dressed as Miss America, complete with sash and a Hollywood-quality makeover, and going to her favorite restaurants and nightclubs. Since the algorithms detected her narcissistic tendencies, the algorithms adjusted the facial expressions and behavior of the people in the restaurants and nightclubs. People seated along Nancy's route stop their conversations to turn their heads and look at her. Their faces radiate admiration and amazement.

Of all the destinations offered, Nancy most enjoys watching herself dance in an upscale nightclub in New York City while wearing her tiara and Miss America sash. In this video prepared by algorithms, she has endless energy and an endless line of people waiting to dance with her. Moreover, the music includes all her favorite songs. Nancy does not know how to dance to some of the songs, but that is not a problem. Advanced artificial intelligence algorithms have scoured recent videos posted online and then determined the ones that were most popular. The algorithms then analyzed the dance moves and applied them to Nancy's avatar. Now Nancy is not only Miss America but also a dancing queen and the life of the nightclub. It is a wonderful performance. People are amazed and awestruck. Many applaud loudly.

Now Nancy is threatened with being terminated from her job because she is spending a lot of time on her smartphone. She has become withdrawn and isolated. Living in a dream world that is made apparently real and exciting by crafty algorithms has become a compulsive obsession for her. Her therapist recommends that she get a brain scan. The device that scans her brain is similar to the one the marketing psychologists used to set up and test Oh Là Là Fashions' website and features.

The brain scan confirms Nancy's therapist's tentative diagnosis. When exposed to images from Oh Là Là Fashions' website, Nancy's brain reacts as strongly as a cocaine addict's brain reacts when the addict looks at cocaine lined out on a table.

Sadly, a team of crack[7] marketing psychologists got to Nancy first. They earned large salaries and bonuses for Oh Là Là Fashions' unprecedented retail performance. In stark contrast, Nancy's therapist is left to deal with a beautiful but devastated and guilt-ridden client who is about to lose her job and go bankrupt. Nancy has very little money, and most health insurance policies do not cover treatment for compulsive shopping. Therefore, Nancy will only have her treatment paid for by her health insurance company if she attempts suicide but fails in that attempt.

* * *

Sally, who has recently become very anxious and irritable, sees a therapist. Sally does not know why she is feeling this way. Her therapist is struggling to make sense of her recent sudden increase in anxiety and irritability. After Sally answers a long and detailed series of questions, the therapist believes that her client's problem may be related to a fear of spiders. However, Sally states that she has increased her pest control service from once every six months to once every six weeks. Although she has not seen any spiders in more than ten months, she believes that her house is infested with spiders. The therapist is starting to think that Sally needs a different psychological diagnosis and perhaps a strong antianxiety medication.

Unknown to either Sally or her therapist is a crafty new app that is making a lot of money by identifying people who are afraid of spiders. Pest control companies are willing to pay a premium for that information.

The clever new app places images of spiders on the edge of the smartphone, outside of Sally's central visual field but within her peripheral vision. Using eye-tracking capability built into the smartphone, the app is able to calibrate Sally's eyes without her awareness that such a thing is being done. The algorithms closely monitor the pattern of her eyes' exploratory saccades when her eyes detect the spider in her peripheral vision. The algorithms can even detect the efforts of her subconscious to restrict her saccades from looking at the spiders.

However, the developers of the app, not wanting to scare potential customers, built in a feature that makes the images of the spiders disappear when the viewer's point of gaze (controlled by the conscious mind) is detected (by the exploratory saccades controlled by the subconscious) as moving to the place where the spiders appeared a few milliseconds earlier.

As a result, Sally's subconscious is sending out an alert to her brain to react to spiders, but Sally's conscious mind is troubled because she cannot recall *seeing* any spiders. As mentioned previously, Sally is suffering increased anxiety and irritability. The app designers and data brokers, on the other hand, are reaping huge profits from this new app that identifies customers who are afraid of spiders and then sells that information to pest controllers.

In this example, you could substitute in Mary's fear of rodents, Bill's fear of snakes, etc., and the results would be much the same. The targeted people will not be able to explain their increased anxiety. However, pest controllers will be happy. Take anyone's fear as an example. Others can profit from those fears. However, there will also be some profound, second-order effects on innocent people.

***

Bill is struggling financially. He can expect to receive advertisements from local casinos showing fourteen people in the preceding week who each won $5,000 or more. Additionally, he can expect to get offers for credit cards; he has been preselected because of his good credit. In actuality, the credit card companies provide him with teaser rates so he will accept the card. The percentage rate of interest will dramatically increase within a few months or when Bill makes one late payment (whichever occurs first).

* * *

Sue, Doug, Sam, and many others like them use search engines to look for information on the Internet in order to broaden their understanding of social and cultural issues, current events, etc. However, the results they receive from the search engines are often biased.[8] In this manner, titans of the Internet can reshape people's perspective of the world. The titans of the Internet manipulate what people see and thereby shape their worldview.

* * *

For poets, the seat of emotion has always been and probably always will be the heart. However, neuroscience has recently allowed us to pinpoint exactly where emotions are processed in the brain. Proverbs 4:23 warns us to guard our hearts. I hope you see that this warning applies just as much (or more) today as it did in ancient times. Modern technology allows others not only to detect and accurately measure our emotions but also to manipulate our emotions if we are not careful. I hope that you are aware of what may soon be possible and that you do take measures to prevent others from deceiving or manipulating you with new technologies.

Legal and Ethical Issues

Because trampling on people's privacy is very profitable, Internet-based corporations have hired lobbyists to convince elected and appointed officials to use a light hand when writing legislation that protects consumer privacy. However, there may be another important dimension. What if elected officials have information that can prevent serious crimes but they say nothing about it?

There arise serious legal and ethical issues if Internet-based corporations have the ability to analyze information as thoroughly as boasted of by Google's CEO ("We know where you are. We know where you've been. We can more or less know what you're thinking about"[1]). That statement was made in 2010, five years before Andreas Lubitz, the copilot of ill-fated Germanwings Flight 9252, researched the Internet to find information about suicide and about cockpit door security.[2] Seeing as Google prides itself on its technological innovation, surely, in the five years since Google's CEO made that boast, Internet-based corporations have made tremendous improvements in their ability to "more or less know what you're thinking about."

Should Internet-based corporations be able to use the information they know about you as a competitive advantage and to boast of their abilities to potential marketers while simultaneously denying they have the ability and eschewing the moral responsibility to save lives when possible? We do not know if Andreas Lubitz used Google or another search engine to look into suicide and cockpit door security. However, we know that providing Internet search engines is a very competitive business. If Lubitz used another search engine, it would likely be one with capabilities similar to Google's. If not, then the search engine company would be out of business.

If Google or other search engine companies can use their ability to know what people are thinking in order to earn huge profits, do they have blood on their hands for remaining silent before 149 people died along with Andreas Lubitz? If one receives advertisements within minutes of viewing a website or purchasing something, then Internet-based corporations certainly could have sent Lufthansa, the parent company of Germanwings, a red-flag message saying that one of its pilots was considering suicide, researching cockpit doors, and researching cockpit security. Many nations have laws that make it illegal for legal entities to have knowledge about imminent illegal actions and to remain silent about it. Therefore, Internet-based corporations could be complicit in the deaths of 149 innocent people. Sometime in the future, courts will have to provide insight into issues like this one. *Minority Report* proposed a scenario wherein the government used technology for the purpose of "precrime" prevention. However, corporations now have a similar capability to understand what people are thinking and what people are likely to do. These companies want to be able to make huge profits by using the new behavioral biometrics. However, should they be held legally liable when they withhold an accurate prediction of imminent public danger posed by terrorists or suicidal murderers? This issue is complicated further when one considers that much of the research and development for behavioral biometrics was funded by taxpayers to protect us from terrorists. Corporations want to profit from these technologies while trying to escape the responsibility of using the information for the purposes that these technologies were originally developed—to protect United States citizens from those who want to harm or kill us.

Until the courts provide the public with guidance on these issues, you will have to make up your own mind. Would you use the products of Internet-based corporations that would seek only to make a profit from you but that would not try to protect you if they had the ability to, say, notify the airline that your pilot or copilot was suicidal and had recently researched cockpit security on the Internet? Depending on how you answer that question, you have several more decisions to make.

What Can You Do?

This book does not advocate giving up your smartphone or going off grid any more than *Unsafe at Any Speed* advocated that people abandon their cars. Anyone concerned about his or her safety, privacy, and smartphone has several options. I encourage you to consider all of these options.

Perhaps the best option, if you want to make a difference, is to spread the word about privacy invasion. Be an advocate for the protection of your own privacy. Get others involved and excited. That is how *Unsafe at Any Speed* led to safer cars and fewer highway fatalities. A nationwide groundswell of people clashed with the titans of the automobile industry, who were once as mighty as the titans of the Internet are today. The consumers ultimately prevailed against the titans back then. Consumers can prevail against today's titans too.

We can prevail against today's titans if we work together and demand change. Talk to friends or text them. Be a trendsetter among your friends by speaking out. If you want to speak out (and you should), then know that it is easy to start a conversion. For some ideas, visit Mozilla's page at https://www.mozilla.org/en-US/privacy/tips/?sample_rate=0.1&snippet_name=4977. Many other people believe, like you do, that their privacy is important, so you are not likely to be viewed as a ranting, raving lunatic. In fact, Pew Research Center reports that most US adults feel that more should be done to protect consumers.[1]

Together we can push back, force changes, and take back control of our private information. This is your chance to help others and to oppose the titans of the Internet who are trampling on our privacy. Decades ago, the automobile titans put profits before safety. Today, the Internet titans put profits before privacy. We can use the marketplace to send a powerful message to the titans of the Internet that privacy is important. You can use your current smartphone a year or two longer than you would otherwise. If enough people delay purchasing or do not purchase a new smartphone, then this will put tremendous financial pressure on smartphone manufactures. If these companies are aware that people are refusing to buy new smartphones because they dislike the invasion of their privacy that new smartphone technologies allow, then the companies will be forced to make smartphones with more features that safeguard user privacy.

If you don't like the fact that Google and other search engines collect your data, then you can use DuckDuckGo (https://duckduckgo.com), an Internet search engine that does not collect user data.[2] If you don't like Facebook's practice of scooping up your information and selling it, then you can use a different social media website. Ello (https://ello.co/wtf/post/privacy) is a social networking site that does not track your activity or sell your information.[3]

If you do not want your smartphone to spy on you with its camera and microphone, then you may cover the camera with a self-adhesive label[4] or a removable label.[5] You may cover the microphone with these labels too, but the label will not completely mute the microphone. Also, it is a little difficult to remove the labels when you want to use your phone. Electrical tape is another material that can be used to cover cameras and microphones. A tab can be made easily by folding over one edge of the tape so it sticks to itself, thus making the tape easy to remove. If you do not want your smart TV spying on you, then you may use self-adhesive labels, removable labels, or electrical tape to cover the TV's microphone.

If enough customers use such methods to prevent being spied on, then manufacturers might respond with smartphones that have easily openable and easily closable flaps to cover the cameras and microphones. This would be an inexpensive hardware solution, one that would allow you to regain control over when the camera and microphone are active. If you want something like a smartphone case with flaps to cover the camera and microphone, then start demanding it.

* * *

Internet-based corporations have been very effective at convincing elected officials that people feel overregulated. If you want laws and regulations to protect your privacy, then you have to demand them—or else lobbyists will tell our elected officials otherwise.

You may contact the President of the United States at http://www.whitehouse.gov/contact.

You may find and contact your two senators at http://www.senate.gov/senators/contact/.

You may find and contact and your representative at http://www.house.gov/representatives/find/.

* * *

Many corporations have an "opt out" link on their websites that allows you to opt out of receiving advertisements tailored to your interests. The link is often not easy to find, but it is worth the effort to try to find it. Often you must scroll down to the bottom of the page to find the "opt out" link, which usually appears in small print.

You may use Digital Advertising Alliance (http://www.aboutads.info/choices/) to opt out of being shown or receiving customized advertisements.

You can hire a company like Reputation.com to help remove your personal data from some corporations' websites (http://www.reputation.com/personal). Reputation.com charges a fee for this service.

Getting Google and Facebook to delete your information is much harder, since your information is at the core of their financial success. These companies are not likely to give up people's information unless they face pressure to do so from many people like you and your friends. Millionaires and billionaires who have made their fortunes by trampling on your privacy rights and selling your information have issued dire warnings that restricting the collection and analysis of users' personal information would destroy the Internet as we know it. This is an exaggeration and a scare tactic.

GLOSSARY

algorithm(s) (ˈal-gə-ri-<u>th</u>əm). A set of steps that are followed in order to solve a mathematical problem or to complete a computer process.[1] The simplest types of algorithms are sets of instructions for a computer to perform a process, such as checking for a software update, or steps instructing machines to performing routine tasks such as printing a document. However, algorithms can be so complex and detailed that they can mimic human behavior. For example, Zeiss, a microscope manufacture, has developed a system that uses "software algorithms for counting, comparison, analysis, and quantification of specific tissue areas or cells of interest."[2] These tasks are normally performed by a trained pathologist. Artificial-intelligence devices are exploring computers to find ways for them to generate their own algorithms and thereby "learn" like a human being does.[3]

Sophisticated software now allows ordinary people, not just experts, to program robots. Robots can "learn with the guide of the user, either through a remote control or through voice commands or simply by showing the automaton tasks, as one would teach a toddler."[4]

anonymizing. A process of separating personal or private information from a name, Social Security Number (SSN), address, etc., so that the identity of the individual is not readily apparent. Sometimes this is a farce. In many cases, it is deceptive. Instead of using a person's name, SSN, address, etc., the person's personal or private information is linked to a digital identifier that is unique (i.e., no one else in the world has the same digital identifier).

anonymous resolution, also known as **anonymous-entity resolution**. is a technology that "scrambles data for security and privacy reasons. The software sifts through data such as names, telephone numbers, addresses, and information from employers to identify individuals listed under different names in separate databases. The software can find information by comparing records in multiple databases, however the information is scrambled using a 'one-way hash function,' which converts a record to a character string that serves as a unique identifier like a fingerprint. Persons being investigated remain anonymous, and agents can isolate particular records without examining any other personal information. A record that has been one-way hashed cannot be 'un-hashed' to reveal information contained in the original record."[5]

artificial intelligence (AI). "the science and engineering of making intelligent machines, especially intelligent computer programs. It is related to the similar task of using computers to understand human intelligence, but AI does not have to confine itself to methods that are biologically observable."[6] Another definition is as follows: "Artificial intelligence, or AI, is a branch of computer science that aims to make computers behave more like humans, with capabilities such as reasoning, learning and planning."[7] Some scientists have come to refer to artificial intelligence as "machine learning" to ease the public's anxiety about robots taking over the world.

autonomous. (ȯ-ˈtä-nə-məs). Existing or acting separately from other things or people.[8] Autonomous programs are often able to act on their own without human intervention except when updating software. Autonomous programs are often designed to mimic expert human behavior when performing tasks. Perhaps seeing is better than just reading. I encourage you to watch the following videos. The German automation company Festo has developed robotic ants that work together to accomplish tasks (https://www.youtube.com/watch?feature=player_embedded&v=FFsMMToxxls) and autonomous bionic butterflies, eMotionButterflies, that are able to fly by themselves or in groups (https://www.youtube.com/watch?v=1gu3z7w4Vc8). In broad terms, the sophistication and ability to perform complex tasks at the lowest level comes from algorithms. At the highest level is artificial intelligence. Autonomous algorithms or autonomous robots are in between simple algorithms and AI.

conscious mind. Many experts believe that the conscious mind handles only 10 percent or less of the mind's thoughts. The subconscious does most of the "thinking," and it forwards the final decision to the conscious mind. This process can be monitored by noting the progression of different brain waves. Marketers often try to avoid pitching advertisements to the conscious mind, as people often screen out and do not act on advertisements pitched this way.

covert attention. This is associated with the subconscious mind as it monitors the peripheral vision. Covert attention is in contrast to overt attention, which is associated with the conscious mind and the object(s) of its attention. Saccades can reveal what are the objects of one's covert attention and thus provide important clues to what the subconscious is interested in. If the subconscious is trying to avoid looking at something with the saccades, then the saccade pattern will be different, thereby providing important clues about a person's possible internal conflicts. If you are interested in learning more, I recommend several scientific articles that are referenced in the notes.[9]

cryptoanalysis. The science and mathematics that enables code breakers to unscramble an encrypted message without having prior knowledge of the encoding key.[10]

cryptanalytic machine. A mechanical or electronic device that enables a code breaker to unscramble encrypted messages.

Dick Tracy watch. Dick Tracy was a police officer portrayed in a comic strip with the same name that debuted in 1931. By 1946, Dick was wearing a "2-Way Wrist Radio, worn as a wristwatch," which may have inspired the later-appearing smartwatches.[11]

fMRI. Magnetic resonance imaging (MRI) is used to show a picture of soft tissue, such as the brain. Functional magnetic resonance imagining (fMRI) can detect areas of the body that are using more oxygen than they normally do. Hence, the word *functional* in fMRI denotes increased functioning of the cells or organs.

hashing. The process of converting numbers or text from an easy to read format to a hard to read, encrypted format. Hashing has its origin in the process used to protect secret messages delivered to members of the government or the military. However, more recently it has been used widely to bypass legal restrictions on personally identifying people and protecting their personal information. The result has been a weakening of privacy laws by technical means. These privacy laws were initially developed to protect secrets.

nucleus accumbens. (ˈn[y]ü-klē-əs\ \-ə-ˈkəm-bənz). The most important thing to know about the nucleus accumbens is that it is involved with rewards, pleasure, and addictions.[12] The location of the nucleus accumbens is more difficult to explain. Also, it is not too important for most people to understand the function of the nucleus accumbens. The nucleus accumbens forms "the floor of the caudal part of the anterior prolongation of the lateral ventricle of the brain."[13] A visual of the location can be found at http://www.health.harvard.edu/newsletter article/how-addiction-hijacks-the-brain.

neuromarketing. (/ˈnjʊərəʊˌmɑːkɪtɪŋ/). "the process of researching the brain patterns of consumers to reveal their responses to particular advertisements and products before developing new advertising campaigns and branding techniques."[14] "The technology is based on a model whereby the major thinking part of human activity (over 90%), including emotion, takes place in the subconscious area that is below the levels of controlled awareness. For this reason, the perception technologists of the market are very tempted to learn the techniques of effective manipulation of the subconscious brain activity. The main reason is to inspire the desired reaction in person's perception as deeply as possible."[15]

overt attention. Attention of the conscious mind. If someone is looking directly at you while you are talking, then you are the object of that person's overt attention. You are the object of that person's conscious mind when his or her gaze axis is fixed on you.

point of gaze; gaze; and gaze axis. These three are similar terms used primarily among neuroscientists who study vision. These things are used to determine what is the primary object a person is looking at or monitoring with his or her conscious mind. The primary object can be determined by monitoring the movement of the eyeball (eye globe). The gaze axis is often associated with the conscious mind and overt attention. The gaze axis is an indication of what is the object of the conscious mind's attention. Computers and algorithms can plot overlays on pictures (heat maps) to reveal precisely where a person is looking.

psychophysiological. "The study of correlations between the mind, behavior, and bodily mechanisms. Also called *physiological psychology*."[16] Although this long word may look intimidating, it expresses a very simple concept. Psychophysiology is the study of why people fidget, sweat, stutter, pace back and forth, etc., when they are nervous.

saccade. (să-kăd', sə-). "A rapid intermittent eye movement, as that which occurs when the eyes fix on one point after another in the visual field."[17] Saccades are often associated with the subconscious, which allows the saccades to monitor peripheral vision. Microsaccades are smaller movements of the eye that allow light to activate fresh rods and cones in the retina while previously discharged rods and cones recharge. Without microsaccades, one's vision would fade when one looked at something.

SST. Steady state topography (SST), a specialized type of EEG (electroencephalograph). SST is a methodology used for observing and measuring human brain activity.[18] Allegedly, SSTs can reveal much about what a person is thinking. Corporations appear to be trying to hide exactly how much they can learn from the human mind with SSTs.

subconscious mind. Some people estimate that more than 90 percent of the mind's thought process occurs in the subconscious. Many decisions are made in the subconscious and are then forwarded to the conscious mind. Saccades are often associated with the subconscious mind. Marketers often try to pitch advertisements to the subconscious mind, as advertisements pitched this way are more effective in influencing or manipulating a person.

Bibliography, References, and Websites

The following are alphabetized using the author's first name, which is followed by the article title and then the URL where the article appears. Please note that *The Economist* does not publish authors' names along with articles; therefore, articles from *The Economist* are alphabetized by their titles—as are other articles that list no author. If the hyperlinks do not launch correctly, cut and paste them into your browser.

Listed by Author's First Name

Adam D. I. Kramer, Jamie E. Guillory, and Jeffrey T. Hancock. "Experimental evidence of massive-scale emotional contagion through social networks." *Proceedings of the National Academy of Sciences of the United States of America.* http://www.pnas.org/content/111/24/8788.full.pdf.

Adam L. Penenberg. "A. K. Pradeep, Mind Reader." *Fast Company.* http://www.fastcompany.com/1772167/ak-pradeep-mind-reader.

Alexei Oreskovic. "Google spells out email scanning practices in new terms of service." *Reuters,* April 14, 2014. http://www.reuters.com/article/2014/04/14/google-email-idUSL2N0N61MT20140414.

Alice Truong. "This Google Glass App Will Detect Your Emotions, Then Relay Them Back to Retailers." *Fast Company.* http://www.fastcompany.com/3027342/fast-feed/this-google-glass-app-will-detect-your-emotions-then-relay-them-back-to-retailers.

Alina Selyukh and Greg Savoy. "Protesters march in Washington against NSA spying." Ed. Peter Cooney. *Reuters,* October 27, 2013. http://www.reuters.com/article/2013/10/27/us-usa-security-protest-idUSBRE99P0B420131027.

Alistair Barr. "Google buys artificial intelligence start-up DeepMind." *USA Today,* January 27, 2014. http://www.usatoday.com/story/tech/2014/01/27/google-deepmind-artificial-intelligence/4943049/. http://www.usatoday.com/story/tech/2014/01/27/google-deepmind-artificial-intelligence/4943049/

———. "New Google Glass Version Coming this Year." *Wall Street Journal.* http://www.wsj.com/video/new-google-glass-version-coming-this-year/B69DDCD7-1C14-4556-B5D9-26487E3CF49D.html (three minutes and thirty-five seconds into the video).

Amber Haq. "This Is Your Brain on Advertising." *Bloomberg Business,* October 8, 2007. http://www.bloomberg.com/bw/stories/2007-10-08/this-is-your-brain-on-advertisingbusinessweek-business-news-stock-market-and-financial-advice.

Andrea Chang. "Facebook to roll out payments feature on Messenger." Accessed March 17, 2015. *Los Angeles Times*, March 17, 2015. http://www.latimes.com/business/technology/la-fi-tn-facebook-messenger-send-money-20150317-story.html?track=rss.

Andrew Griffin. "Samsung's new smart TV policy allows company to listen in on users." Accessed February 9, 2015. *The Independent*. http://www.independent.co.uk/life-style/gadgets-and-tech/news/samsungs-new-smart-tv-policy-allows-company-to-listen-in-on-users-10033012.html?icn=puff-13.

Ann Marie Seward Barry. *Visual Intelligence, Perception, Image, and Manipulation in Visual Communication*. Albany: State University of New York Press, 1997.

Anton Troianovski. "Phone Firms Sell Data on Customers." *Wall Street Journal*. May 21, 2013, http://www.wsj.com/articles/SB10001424127887323463704578497153556847658.

Aries Poon. "Hon Hai Says It Sold Some Display Patents to Google." *Wall Street Journal*. http://online.wsj.com/article/BT-CO-20130823-704061.html.

Ben Griffin. "ENISA: Smartphones a 'goldmine of sensitive and personal information.'" *Dennis Publishing*. http://www.knowyourmobile.com/products/11341/enisa-smartphones-goldmine-sensitive-and-personal-information.

Ben Mezrich. *The Accidental Billionaires: The Founding of Facebook: A Tale of Sex, Money, Genius, and Betrayal*. New York: Doubleday, 2009.

Ben Rossi. "Facebook DOES collect the text you decided against posting." *Information Age*. http://www.information-age.com/technology/information-management/123459286/facebook-does-collect-text-you-decided-against-posting.

Bill Meyer. "How many Americans' Social Security numbers were officially duplicated for Pacific islanders?" August 18, 2009. *Cleveland.com* (Plain Dealer Publishing Co.). http://www.cleveland.com/nation/index.ssf/2009/08/how_many_americans_social_secu.html.

Bill Rigby. "Exclusive: Six percent of U.S. adults plan to buy Apple Watch—Reuters/Ipsos poll." *Reuters,* April 15, 2015. http://www.reuters.com/article/2015/04/15/us-apple-watch-idUSKBN0N628820150415.

Bradley Voytek. "Mapping the San Franciscome." January 9, 2012. *Uber.* http://blog.uber.com/2012/01/09/uberdata-san-franciscomics/.

Brian Knutson. "Visualizing Desire." Uploaded January 22, 2009; accessed March 17, 2015. https://www.youtube.com/watch?v=CUK8D-kX0fE.

Brody Mullins, Rolfe Winkler, and Brent Kendall. "Inside the U.S. Antitrust Probe of Google." Last modified March 19, 2015. *Wall Street Journal.* http://www.wsj.com/articles/inside-the-u-s-antitrust-probe-of-google-1426793274.

Bruce Schneier. *Data and Goliath: The Hidden Battle to Collect Your Data and Control You.* New York: W. W. Norton, 2015.

Carl P. Brown. "Social Security Numbers—Use & Abuse" (originally appeared in *Bankers' Hotline* 1 [1], January

1990). *Bankers Online*. http://www.bankersonline.com/articles/bhv01n01/bhv01n01a6.html.

Carl Zimmer. "Secrets of the Brain." *National Geographic* 225, no. 2 (February 2014).

Carolyn Giardina. "'Furious 7' and How Peter Jackson's Weta Created Digital Paul Walker." March 25, 2015. *The Hollywood Reporter*. http://www.hollywoodreporter.com/behind-screen/furious-7-how-peter-jacksons-784157.

Charlie Osborne. "Google launches quantum processor, artificial intelligence project." *ZDNet*. http://www.zdnet.com/article/google-launches-quantum-processor-artificial-intelligence-project/.

Charles Duhigg. "How Companies Learn Your Secrets." *New York Times,* February 19, 2012. http://www.nytimes.com/2012/02/19/magazine/shopping-habits.html?pagewanted=all&_r=0.

Chris Crum. "Google Outlines What It's Doing to Protect Your Data from the Government." January 2013. *WebProNews*. http://www.webpronews.com/google-outlines-what-its-doing-to-protect-your-data-from-the-government-2013-01.

Chris Foresman. "Apple breaks ground on mammoth Oregon data center." October 2012. *Ars Technica*. http://arstechnica.com/apple/2012/10/apple-breaks-ground-on-mammoth-colossal-gargantuan-oregon-data-center/.

Chris Wood. "Google eye tracking unlock patent revealed." August 8, 2012, *Gizmag*. http://www.gizmag.com/google-eye-track-unlock-patent/23637.

Chriss W. Street. "i-Watch Could Let Managers Snoop on Employees." April 6, 2015. *Breitbart.* http://www.breitbart.com/california/2015/04/06/i-watch-could-let-managers-snoop-on-employees/.

Claudia Assis. "The list of electric-car projects to rival Tesla is getting longer." February 20, 2012. *Market Watch.* http://www.marketwatch.com/story/the-list-of-electric-car-projects-to-rival-tesla-is-getting-longer-2015-02-20?siteid=rss&rss=1.

Coby Ben-Simhon. "Queen of broken hearts." August 6, 2010 *Haaretz.* http://www.haaretz.com/weekend/magazine/queen-of-broken-hearts-1.306416.

Daisuke Wakabayashi. "Apple to Build Data Command Center in Arizona." *Wall Street Journal,* February 2, 2015. http://blogs.wsj.com/digits/2015/02/02/apple-to-build-data-command-center-in-arizona/.

Dana Mattioli. "On Orbitz, Mac Users Steered to Pricier Hotels." August 23, 2012. *Wall Street Journal.* http://www.wsj.com/articles/SB10001424052702304458604577488822667325882.

Daniel Eran Dilger. "First Look: Apple's iCloud data center site in Reno, Nevada." March 9, 2013. *Apple Insider.* http://appleinsider.com/articles/13/03/09/first-look-apples-icloud-data-center-site-in-reno-nevada.

Daniel Gardner. *Risk: The Science and Politics of Fear.* London: Virgin Books, 2009.

———. *Science of Fear: How the Culture of Fear Manipulates People's Brains.* London: Virgin Books, 2009.

"Data Fusion: Information of the World Unite! Privacy in an Age of Terabytes and Terror." *Scientific American* 299, no. 3 (September 2008).

David Alton Clark. "Apple: About to Show You the iMoney?" June 7, 2013. *Seeking Alpha*. http://seekingalpha.com/article/1488192-apple-about-to-show-you-the-imoney.

David Finn. "Health Information = A Hacker's Gold Mine." June 18, 2012. *Symantec*. http://www.symantec.com/connect/blogs/health-information-hacker-s-gold-mine.

David Talbot. "Startup Gets Computers to Read Faces, Seeks Purpose Beyond Ads." October 28, 2013. *MIT Technology Review*. http://www.technologyreview.com/news/519656/startup-gets-computers-to-read-faces-seeks-purpose-beyond-ads/.

Devlin Barrett. "CIA Aided Program to Spy on U.S. Cellphones." *Wall Street Journal,* March 10, 2015. http://www.wsj.com/articles/cia-gave-justice-department-secret-phone-scanning-technology-1426009924?mod=trending_now_10.

Dirk Hanson. "The Nucleus Accumbens." February 2010. *Addiction Inbox Blog*. http://addiction-dirkh.blogspot.com/2010/02/nucleus-accumbens.html.

Dominic Rushe. "Facebook IPO sees Winklevoss twins heading for $300m fortune." *The Guardian,* February 2, 2012. http://www.theguardian.com/technology/2012/feb/02/facebook-ipo-winklevoss-300m-fortune.

Douglas van Praet. *Unconscious Branding: How Neuroscience Can Empower (and Inspire) Marketing.* New York: Palgrave Macmillan, 2012.

Drew Harwell. "Pizza Hut wants to read your mind." *Washington Post,* December 1, 2014. http://www.washingtonpost.com/blogs/the-switch/wp/2014/12/01/pizza-hut-wants-to-read-your-mind/.

Dyan Machan. "The New Information Goldmine." *Wall Street Journal.* http://online.wsj.com/article/SB125071202052143965.html.

Ed Felten. "Does Hashing Make Data 'Anonymous'?" April 22, 2012. *Tech@FTC Blog.* https://techatftc.wordpress.com/2012/04/22/does-hashing-make-data-anonymous.

Edna B. Foa and Michael J. Kozak. "Emotional Processing of Fear: Exposure to Corrective Information." *Psychological Bulletin* 99, no. 1 (1986): 20–35. http://www.personal.kent.edu/~dfresco/CBT_Readings/foa_&_kozak_1986.pdf.

Edna B. Foa, Elizabeth A. Hembree, and Barbara Olawov Rothbaum. *Prolonged Exposure Therapy for PTSD, Emotional Processing of Traumatic Events, Therapist Guide (Treatments that Work).* New York: Oxford University Press, 2007.

Elaine Ganley, Angela Charlton, and Frank Jordans. "Germanwings Co-Pilot Researched Suicide Methods, Cockpit Security: Prosecutors." *Huffington Post,* April 2, 2015. http://www.huffingtonpost.com/2015/04/02/andreas-lubitz-suicide-methods_n_6992420.html.

Ellen Gamerman. "When the Art Is Watching You." *Wall Street Journal.* http://www.wsj.com/articles/when-the-art-is-watching-you-1418338759?KEYWORDS=data+mining+medical+information.

Ellen Shapiro. Ed. Kelly Knauer. *The 100 Most Influential People Who Never Lived.* New York: Time Books, 2013.

Emily Steel. "Companies scramble for consumer data." *Financial Times.* http://www.ft.com/intl/cms/s/0/f0b6edc0-d342-11e2-b3ff-00144feab7de.html#axzz2W71Wg6o0 (subscription required).

Elizabeth Dwoskin. "What Secrets Your Phone Is Sharing about You." *Wall Street Journal.* http://online.wsj.com/news/articles/SB10001424052702303453004579290632128929194?mod=trending_now_1.

Elizabeth Dwoskin and Evelyn M. Rusli. "The Technology that Unmasks Your Hidden Emotions." *Wall Street Journal.* http://www.wsj.com/articles/startups-see-your-face-unmask-your-emotions-1422472398?mod=LS1.

Erik Sherman. "How Men Look at Women: One Online Marketer's Surprise Answer." *CBS News MoneyWatch.* http://www.cbsnews.com/news/how-men-look-at-women-one-online-marketers-surprise-answer/.

Erica Warp. "NeuroDisco." http://ericawarp.com/?projects=neurodisco.

Everett Rosenfeld. "Apple announces $2B global command center in Arizona." *CNBC.* http://www.cnbc.com/id/102389945.

"Facial Action Coding System." http://facialactioncoding.com/.

Geoffrey Smith. "Apple's Crazy-Expensive New Data Centers Will Be Totally Green." *Time*. http://time.com/3718616/apple-data-centers-green/.

George Lakoff and Mark Johnson. *Philosophy in the Flesh: The Embodied Mind and Its Challenge to Western Thought.* New York: Basic Books, 1999.

Gerald Zaltman. *How the Consumers Think: Essential Insight into the Mind of the Market.* Boston: Harvard Business School Publishing, 2003.

Greg Masters. "Rite Aid to pay $1 million fine for HIPAA violation." *SC Magazine*. http://www.scmagazine.com/rite-aid-to-pay-1-million-fine-for-hipaa-violation/article/175729/.

Gus Lubin, Kim Bhasin, and Shlomo Sprung. "16 Heatmaps That Reveal Exactly Where People Look." May 2012. *Business Insider*. http://www.businessinsider.com/eye-tracking-heatmaps-2012-5?op=1.

Guy Grimland. "Israel startup uses behavioral science to identify terrorists." *Haaretz*. http://www.haaretz.com/print-edition/business/israel-startup-uses-behavioral-science-to-identify-terrorists-1.245470.

Hamish Pringle. "Why Emotional Messages Beat Rational Ones." *Advertising Age*. http://adage.com/article/cmo-strategy/emotional-messages-beat-rational/134920/2015.

Hamish Pringle and Peter Field. *Brand Immortality.* Philadelphia: Kogan Page, 2008.

Hannah Devlin. "What Is Functional Magnetic Resonance Imaging (fMRI)?" *Psych Central*. http://psychcentral.

com/lib/what-is-functional-magnetic-resonance-imaging-fmri/0001056.

Heather Green. "The Information Gold Mine." 1999. *Business Week.* http://www.businessweek.com/1999/99_30/b3639030.htm.

Igor Bobic. "Edward Snowden Explains How the Government Can Get Your 'Dick Pic' during Interview with John Oliver." *Huffington Post,* April 6, 2015. http://www.huffingtonpost.com/2015/04/06/edward-snowden-john-oliver-dick-pic_n_7008330.html.

Jack M. Wedam. US patent 4,654,647, "Finger actuated electronic control apparatus." http://patft.uspto.gov/netacgi/nph-Parser?Sect1=PTO1&Sect2=HITOFF&d=PALL&p=1&u=%2Fnetahtml%2FPTO%2Fsrchnum.htm&r=1&f=G&l=50&s1=4,654,647.PN.&OS=PN/4,654,647&RS=PN/4,654,647.

James B. Twitchell. *Twenty Ads that Shook the World.* New York: Crown Publishing, 2000.

James S. Coleman. *Introduction to Mathematical Sociology.* New York: Free Press of Glencoe/Macmillan, 1964.

Jean Kilbourne. *Deadly Persuasions.* New York: The Free Press, 1999.

Jeffrey Rosen. "The Brain on the Stand." *New York Times Magazine,* March 11, 2007. http://www.nytimes.com/2007/03/11/magazine/11Neurolaw.t.html?pagewanted=6&_r=0.

Jenna Wortham. "Apple Buys a Start-Up for Its Voice Technology." *New York Times,* April 29, 2010. http://www.nytimes.com/2010/04/29/technology/29apple.html?_r=0.

Jianguo Li, Eric Li, Yurong Chen, and Lin Xu. "Visual 3D Modeling from Images and Videos." *Intel Labs* (China). https://16cbeb23-a-62cb3a1a-s-sites.googlegroups.com/site/leeplus/3DTR.pdf?attachauth=ANoY7cq2WWmGGpnGHFNxYdWzyEeNb4aOj5xhkQUsg43k8OgtU4H7f97QrXnkmXiJ5HODn9y0k-zrO-poJsp3ayh4OP7z0P6xLWP4GEUswVosOowG7f2abS6a4cy0653qjcD8rVieTDcIbefc0frf-LqzAhwhbHSjbZ_bwl9-BATeE50DUcvCB3JPPCJDVQueImLKY5p4KIIU&attredirects=0 and.

Jianguo Li, Eric Li, Yurong Chen, Lin Xu, and Yimin Zhang. "Bundled Depth-Map Merging for Multiview Stereo." https://docs.google.com/file/d/0B8_ZlPz5Dh7mUHlCM0VCLXhUYnEzV1lGUDFhTkxVUQ/edit?pli=1.

Joab Jackson. "IBM Watson Vanquishes Human Jeopardy Foes." *PCWorld.* http://www.pcworld.com/article/219893/ibm_watson_vanquishes_human_jeopardy_foes.html.

Jochen Laubrock, Ralf Engbert, Martin Rolfs, and Reinhold Kliegl. "Commentary: Microsaccades Are an Index of Covert Attention Commentary on Horowitz, Fine, Fencsik, Yurgenson, and Wolfe (2007)." *Psychological Science* 18, no. 4. http://lpp.psycho.univ-paris5.fr/pdf/PapersMR/2007/Laubrock-18-2007-364-6%20discussion%20367-8.pdf.

———. "Microsaccades Are an Index of Covert Attention: Commentary on Horowitz, Fine, Fencsik, Yurgenson, and Wolfe (2007)." *Psychological Science.* http://pss.sagepub.com/content/18/4/364.extract.

Joel Rosenblatt. "Google Wants E-Mail Scanning Information Blocked." *Bloomberg,* March 14, 2014. http://www.bloomberg.com/news/2014-03-14/google-wants-e-mail-scanning-information-blocked.html.

John Markoff. "Google's Next Phase in Driverless Cars: No Steering Wheel or Brake Pedals." *New York Times,* May 28, 2015. http://www.nytimes.com/2014/05/28/technology/googles-next-phase-in-driverless-cars-no-brakes-or-steering-wheel.html?_r=0.

John McCarthy. "What Is Artificial Intelligence?" *Stanford University.* http://www-formal.stanford.edu/jmc/whatisai/node1.html.

Jon Buys. "DuckDuckGo: A New Search Engine Built from Open Source." *OStatic.* http://ostatic.com/blog/duckduckgo-a-new-search-engine-built-from-open-source.

Jon Phillips. "The 10 most likely sensors in a 10-sensor Apple smartwatch." http://www.techhive.com/article/2366126/the-10-most-likely-sensors-in-a-10-sensor-apple-smartwatch.html.

Jonah Lehrer. *How We Decide.* New York: Houghton Miffin Harcourt, 2009.

Jonathan R. Cole. "Paul F. Lazarsfeld: His Scholarly Journey." *Columbia University.* http://www.columbia.edu/cu/univprof/jcole/_pdf/2004Lazarsfeld.pdf.

Jonathan Stempel. "Google won't face email privacy class action." *Reuters,* March 19, 2014. http://www.reuters.com/article/2014/03/19/us-google-gmail-lawsuit-idUSBREA2I13G20140319.

Jordan Robertson. "Your Medical Records Are for Sale." August 8, 2013. *Bloomberg*. http://www.bloomberg.com/bw/articles/2013-08-08/your-medical-records-are-for-sale.

Jorge Otero-Millan et al. "Saccades and microsaccades during visual fixation, exploration, and search: Foundations for a common saccadic generator." *Journal of Vision*. http://www.journalofvision.org/content/8/14/21.long.

Joseph LeDoux. *The Emotional Brain: The Mysterious Understanding of Emotions*. New York: Simon and Schuster, 1996.

Joseph Walker. "Can a Smartphone Tell if You're Depressed?" *Wall Street Journal*. http://www.wsj.com/articles/can-a-smartphone-tell-if-youre-depressed-1420499238?mod=WSJ_hpp_MIDDLENexttoWhatsNewsThird.

Josh Gerstein and Stephanie Simon. "Who watches the watchers? Big Data goes unchecked." May 2014. *Politico*. http://www.politico.com/story/2014/05/big-data-beyond-the-nsa-106653.html.

Julia Angwin. "The Web's New Gold Mine: Your Secrets." *Wall Street Journal*. http://online.wsj.com/article/SB10001424052748703940904575395073512989404.html?mod=WSJ_WhatTheyKnow2010_WhatsNews_4_2_Left.

Julia Angwin and Jeremy Singer-Vine. "Selling You on Facebook." *Wall Street Journal*. http://online.wsj.com/article/SB10001424052702303330250457732774009046230.html?mod=WSJ_WhatTheyKnowPrivacy_MIDDLETopMiniLeadStory.

Julia Fioretti. "Facebook 'tramples European privacy law': Belgian watchdog." *Reuters,* May 15, 2015. http://www.reuters.com/article/2015/05/15/us-facebook-eu-privacy-idUSKBN0OO0XW20150515.

Julian Hattem. "Facebook claims 'a bug' made it track nonusers." *The Hill.* http://thehill.com/policy/technology/238399-facebook-claims-a-bug-made-it-track-people-not-on-facebook.

Julian Ryall. "Robots provide a personal touch at Japanese bank." *The Telegraph.* http://www.telegraph.co.uk/news/worldnews/asia/japan/11384391/Robots-provide-a-personal-touch-at-Japanese-bank.html.

Kashmir Hill. "How Target Figured Out a Teen Girl Was Pregnant before Her Father Did." *Forbes,* February 2, 2016. http://www.forbes.com/sites/kashmirhill/2012/02/16/how-target-figured-out-a-teen-girl-was-pregnant-before-her-father-did/.

———. "Blueprints of NSA's Ridiculously Expensive Data Center in Utah Suggest It Holds Less Info than Thought." *Forbes,* July 24, 2013. http://www.forbes.com/sites/kashmirhill/2013/07/24/blueprints-of-nsa-data-center-in-utah-suggest-its-storage-capacity-is-less-impressive-than-thought/.

Katharine A. Kaplan. "Facemash Creator Survives Ad Board." *The Harvard Crimson* (November 19, 2003). http://www.thecrimson.com/article/2003/11/19/facemash-creator-survives-ad-board-the/.

Kathryn Hauser. "Retailers Using Science to Shape Shopping Experience." *WBZ-TV,* December 9,

2014. http://boston.cbslocal.com/2014/12/09/retailers-using-science-to-shape-shopping-experience/.

Keach Hagey, Shalini Ramachandran, and Daisuke Wakabayashi. "Apple Plans Web TV Service in Fall." *Wall Street Journal.* http://www.wsj.com/articles/apple-in-talks-to-launch-online-tv-service-1426555611?mod=WSJ_hp_LEFTTopStories.

Keith J. Kaplan, MD. "ZEISS Axio Scan.Z1 Advanced Digital Imaging System Improves Pathology Research." *Digital Pathology Blog.* http://tissuepathology.com/author/justinclarkro/.

Kerri Smith. "Brain makes decisions before you even know it." *Nature* (2008). http://www.nature.com/news/2008/080411/full/news.2008.751.html.

Kris Van Cleave. "Eye-tracking technology helps marketers and medical professionals alike." *WJLA,* May 2012. http://www.wjla.com/articles/2012/05/eye-tracking-technology-helps-marketers-and-medical-professionals-alike-75702.html.

Kyle Smith. "Google controls what we buy, the news we read—and Obama's policies." *New York Post,* March 28, 2015. http://nypost.com/2015/03/28/google-controls-what-we-buy-the-news-we-read-and-obamas-policies/.

Lawrence R. Samuel. *Freud on Madison Avenue.* Philadelphia: University of Pennsylvania Press, 2010.

Lisa A. Williams and Eliza Bliss-Moreau. "Your smartphone is looking at you—but can it read your emotions?" March 11, 2014. Accessed January 17, 2015. *The Conversation.* http://theconversation.com/

your-smartphone-is-looking-at-you-but-can-it-read-your-emotions-23908.

Lori Andrews. *I Know Who You Are and I Saw What You Did.* New York: Simon and Schuster, 2010. Kindle edition. (See p. 5, location 209 of 7,414.)

Lorraine Luk. "Foxconn Sells Communications Technology Patents to Google." *Wall Street Journal.* http://online.wsj.com/news/articlesSB10001424052702304788404579523051086783712?mod=WSJ_hps_MIDDLE_Video_Top&mg=reno64-wsj.

Lydia Sloan Cline. *3D Printing with Autodesk 123D, Tinkercad, and Makerbot.* New York: McGraw Hill, 2015.

Malcolm Gladwell. "The Naked Face." *Gladwell.com.* http://gladwell.com/the-naked-face/.

Manda Mahoney. "The Subconscious Mind of the Consumer (And How to Reach It)." Accessed April 24, 2013. *Harvard Business School Working Knowledge.* http://hbswk.hbs.edu/item/3246.html.

Marc Andrews, Matthijs van Leeuwen, and Rick van Baaren. *Hidden Persuasion: 33 Psychological Influences Techniques in Advertising.* Amsterdam: BIS Publishers, 2013.

Marcy Wilder. "Alaska Medicaid Settles HIPAA Security Rule Violations for $1.7 Million." June 2012. *Hogan Lovells.* http://www.hldataprotection.com/2012/06/articles/health-privacy-hipaa/alaska-medicaid-settles-hipaa-security-rule-violations-for-17-million/.

Margaret Talbot. "Duped." *New Yorker,* July 2, 2007. http://www.newyorker.com/magazine/2007/07/02/duped.

Marie Mawad. "Apple to Spend $1.9 Billion Building Two Europe Data Centers." February 2, 2015. *Bloomberg*. http://www.bloomberg.com/news/articles/2015-02-23/apple-to-build-1-9-billion-data-centers-in-denmark-ireland.

Marisa Carrasco. "Visual attention: The past 25 years." *Vision Research* 51, no. 13 (2011): 1,484–1,525. http://www.sciencedirect.com/science/article/pii/S0042698911001544.

"Mark Zuckerberg." *Brainy Quote*. Accessed April 10, 2015. http://www.brainyquote.com/quotes/quotes/m/markzucker453439.html#pch6Q7bjXPiFm2Cc.99.

Mark Hosenball. "NSA chief says Snowden leaked up to 200,000 secret documents." *Reuters*, November 14, 2013. http://www.reuters.com/article/2013/11/14/us-usa-security-nsa-idUSBRE9AD19B20131114.

Mark Prigg. "Glass without the glasses: Google patents smart contact lens system with a camera built in." *Daily Mail*. http://www.dailymail.co.uk/sciencetech/article-2604543/Glass-without-glasses-Google-patents-smart-contact-lens-CAMERA-built-in.html.

———. "Now that's a REAL iComputer: Samsung unveils eye and blink controlled machine to help the disabled get online." *Daily Mail*. http://www.dailymail.co.uk/sciencetech/article-2850843/Samsung-unveils-EyeCan-machine-help-disabled-online.html#ixzz3KDpWzgpK.

Martin Lindstrom. *Buyology*. New York: Crown Business, 2010.

Martin Rolfs, Ralf Engber, and Reinhold Kliegl. "Crossmodal coupling of oculomotor control and spatial attention in vision and audition." *Experimental Brain Research* 166 (2005): 427. doi: 10.1007/s00221-005-2382-y. http://lpp.psycho.univ-paris5.fr/pdf/PapersMR/2005/Rolfs-166-2005-427-439.pdf.

Mary Carmichael. "Neuromarketing: Is it coming to a lab near you?" *PBS*. http://www.pbs.org/wgbh/pages/frontline/shows/persuaders/etc/neuro.html.

Mary Madden. "Public Perceptions of Privacy and Security in the Post-Snowden Era." *Pew Research Center*. http://pewrsr.ch/1wlEbYR.

Mary Wollstonecraft Shelley. *Frankenstein; or, The Modern Prometheus*. Garden City, NY: Halcyon House, 2010.

Matt Cover. "Government Gave 4,317 Aliens 2 Social Security Numbers a Piece." *CNS News*. http://cnsnews.com/news/article/government-gave-4317-aliens-2-social-security-numbers-piece.

Matt Rosoff. "Is Google A Monopoly? 'We're in that Area,' Admits Schmidt." *Business Insider,* September 21, 2011. http://articles.businessinsider.com/2011-09-21/tech/30183638_1_monopoly-web-browser-market-microsoft.

Matthew J. Salganik, Dimitri Fazito, Neilane Bertoni, Alexandre H. Abdo, Maeve B. Mello, and Francisco I. Bastos. "Assessing Network Scale-up Estimates for Groups Most at Risk of HIV/AIDS: Evidence From a Multiple-Method Study of Heavy Drug Users in Curitiba, Brazil." *American Journal of Epidemiology* 174, no. 10. doi: 10.1093/aje/kwr246. http://www.princeton.edu/~mjs3/salganik_assessing_2011.pdf.

Matthew J. Salganik. "How Do Social Patterns Emerge from the Actions and Interactions of Individuals?" *Princeton University.* http://www.princeton.edu/sociology/faculty/salganik/.

Mia De Graaf. "Head of Google's secretive X lab defends Glass headset against privacy campaigners—and explains why engineers now wear fluffy socks to work." *Daily Mail.* http://www.dailymail.co.uk/sciencetech/article-2999644/Head-Google-s-secretive-X-lab-defends-firm-s-Glass-headset-against-privacy-campaigners-explains-engineers-wear-fluffy-socks-work.html.

Michael Fertik. "Federal Trade Commission Ponders Digital Privacy." December 12, 2009. Accessed February 21, 2015. *Michael Fertik.* http://michaelfertik.com/news/federal-trade-commission-ponders-digital-privacy/.

Michal Kosinski, David Stillwell, and Thore Graepel. "Private traits and attributes are predictable from digital records of human behavior." *Proceedings of the National Academy of Science of the United States of America* 110, no. 15: 5,802–05. doi: 10.1073. http://www.pnas.org/content/110/15/5802.full.

Natasha Lomas. "Today in Creepy Privacy Policies, Samsung's Eavesdropping TV." February 8, 2015. *TechCrunch.* http://techcrunch.com/2015/02/08/telescreen/.

Natasha Singer. "A Vault for Taking Charge of Your Online Life." *New York Times,* December 9, 2012. http://www.nytimes.com/2012/12/09/business/company-envisions-vaults-for-personal-data.html?pagewanted=all.

Nate Anderson. "Why Google keeps your data forever, tracks you with ads." March 2010. *Ars Technica*. http://arstechnica.com/tech-policy/2010/03/google-keeps-your-data-to-learn-from-good-guys-fight-off-bad-guys/.

Nicholas Carlson. "At Last—The Full Story of How Facebook Was Founded." *Business Insider*. http://www.businessinsider.com/how-facebook-was-founded-2010-3?op=1.

———. "In 2004, Mark Zuckerberg Broke into a Facebook User's Private Email Account." *Business Insider*. http://www.businessinsider.com/how-mark-zuckerberg-hacked-into-the-harvard-crimson-2010-3.

Nick Pickels. "Google Glass: Orwellian surveillance with fluffier branding." *The Telegraph*. http://www.telegraph.co.uk/technology/google/9939933/Google-Glass-Orwellian-surveillance-with-fluffier-branding.html.

Nick Saint. "Google CEO: 'We Know Where You Are. We Know Where You've Been. We Can More or Less Know What You're Thinking About.'" *Business Insider*. http://www.businessinsider.com/eric-schmidt-we-know-where-you-are-we-know-where-youve-been-we-can-more-or-less-know-what-youre-thinking-about-2010-10.

Patrick Renvoise and Christopher Morin. *Neuromarketing: Understanding the "Buy Buttons" in Your Customer's Brain*. Dallas: Thomas Nelson, 2007.

Paul Ekman. *Emotions Revealed*. New York: St. Martin's Press, 2007.

Peter Holley. "Bill Gates on dangers of artificial intelligence: 'I don't understand why some people are not concerned.'"

Washington Post, January 28, 2015. http://www.washingtonpost.com/blogs/the-switch/wp/2015/01/28/bill-gates-on-dangers-of-artificial-intelligence-dont-understand-why-some-people-are-not-concerned/.

Phillip Bonacich and Philp Lu. *Introduction to Mathematical Sociology.* Princeton: Princeton University Press, 2012.

Ralf Engbert and Kliegl Reinhold. "Microsaccades uncover the orientation of covert attention." *Vision Research* 43, no. 9 (2003): 1,035–45. http://www.sciencedirect.com/science/article/pii/S0042698903000841.

Reed Albergotti. "Facebook Buys Voice-Recognition Startup." *Wall Street Journal.* http://www.wsj.com/articles/facebook-buys-voice-recognition-startup-1420496634?mod=WSJ_hp_LEFTWhatsNewsCollection.

Ren Lu. "Making a Bayesian Model to Infer Uber Rider Destinations." *Uber.* http://blog.uber.com/passenger-destinations.

Rick Jervis. "Sirius founder envisions world of cyber clones, tech med." *USA Today,* March 15, 2015. http://www.usatoday.com/story/tech/2015/03/15/sxsw-rothblatt-cyber-clones-keynote/24816839/.

Rich Miller. "Apple Buys California Data Center." February 27, 2006. *Data Center Knowledge.* http://www.datacenterknowledge.com/archives/2006/02/27/apple-buys-california-data-center/.

———. "Apple Adding Data Center in Silicon Valley." May 18, 2011. *Data Center Knowledge.* http://www.

datacenterknowledge.com/archives/2011/05/18/apple-adding-data-center-in-silicon-valley/.

———. "Apple Confirms Plans for Oregon Data Center." February 2, 2012. *Data Center Knowledge.* http://www.datacenterknowledge.com/archives/2012/02/21/apple-confirms-plans-for-oregon-data-center/.

Robert Heath. *Seducing the Subconscious: The Psychology of Emotional Influence in Advertising.* Malden, MA: Wiley Publishing, 2012.

Robin Respaut and Lucas Iberico Lozada. "'Slicing and dicing': How some U.S. firms could win big in 2016 elections." *Reuters,* April 14, 2015. http://www.reuters.com/article/2015/04/14/us-usa-election-data-idUSKBN0N509O20150414.

Robin Sidel and Daisuke Wakabayashi. "Apple Pay Stung by Low-Tech Fraudsters." *Wall Street Journal.* http://www.wsj.com/articles/apple-pay-stung-bylow-techfraudsters-1425603036?mod=WSJ_hp_LEFTTopStories.

Roger Dooley. *Brainfluence.* Hoboken: John Wiley and Son, 2012.

———. "Neuromarketing: Where Brain Science and Marketing Meet." *Neuroscience Marketing.* http://www.neurosciencemarketing.com/blog.

Rohan Joseph D'Sa. "Emotion recognition of speech signals." *RTNS.* http://www.rtns.org/rohan/Emotion%20recognition%20of%20speech%20%20signals.htm.

Sam Gosling. *Snoop: What Your Stuff Says about You.* New York: Basic Books, 2008.

Sara Boboltz. "5 Extremely Private Things Your iPhone Knows about You." *Huffington Post,* March 19, 2015. http://www.huffingtonpost.com/2015/03/19/iphone-legal-facts_n_6787876.html?ir=Weird+News&ncid=fcbklnkushpmg00000022.

Sara Silverstein. "These Animated Charts Tell You Everything about Uber Prices in 21 Cities." October 2014. *Business Insider.* http://www.businessinsider.com/uber-vs-taxi-pricing-by-city-2014-10.

Sara Smyth. "Toll of social media on girls' mental health: Sexualised images fuelling rise in anxiety among pupils aged 11 to 13." *Daily Mail.* http://www.dailymail.co.uk/news/article-3046222/Toll-social-media-girls-mental-health-Sexualised-images-fuelling-rise-anxiety-pupils-aged-11-13.html.

Scott Austin, Chris Canipe, and Sarah Slobin. "The Billion Dollar Startup Club." Accessed February 19, 2015. *Wall Street Journal.* http://graphics.wsj.com/billion-dollar-club/?co=Uber.

Scott Thurm and Yukari Iwatani Kane. "Your Apps Are Watching You." *Wall Street Journal.* http://online.wsj.com/article/SB10001424052748704694004576020083703574602.html?mod=WSJ_WhatTheyKnow2010_WhatsNews_4_2_Right_Summaries.

Septime Meunier. "Health checks by smartphone raise privacy fears." *Yahoo! News.* http://news.yahoo.com/health-checks-smartphone-raise-privacy-fears-080635403.html;_ylt=AwrSyCOdk_hUcAsATy7QtDMD.

Shelley Carson. "The Creative Mind." *Scientific American Mind* (2014).

Simon Singh. *The Code Book*. New York: Anchor Books, 1999.

Steve Rosenbush. "Facebook Tests Software to Track Your Cursor on Screen." Accessed May 22, 2014. *Wall Street Journal*, October 30, 2013. http://blogs.wsj.com/cio/2013/10/30/facebook-considers-vast-increase-in-data-collection/.

Steven Walden and Qaalfa Dibeehi. "Why we must measure emotion." *Research* (August 2010). http://www.research-live.com/magazine/why-we-must-measure-emotion/4003434.article.

Susan Jones. "IG Audit: 6.5 Million People with Active Social Security Numbers Are 112 or Older." *CNS News*. http://www.cnsnews.com/news/article/susan-jones/ig-audit-65-million-people-active-social-security-numbers-are-112-or-older.

Susana Martinez-Conde, Jorge Otero-Millan, and Stephen L. Macknik. "The impact of microsaccades on vision: toward a unified theory of saccadic function." *Nature Reviews Neuroscience* 14 (2013): 83–96. doi:10.1038/nrn3405. http://www.nature.com/nrn/journal/v14/n2/full/nrn3405.html (subscription required).

Susana Martinez-Conde and Stephen L. Macknik. "Mystery Solved." *Scientific American Mind* 22, no. 5 (November–December 2011): 54 (subscription required).

———. "Shifting Focus." *Scientific American Mind* 22, no. 5 (November–December 2011): 48–55 (subscription required).

———. "Windows of the Mind." *Scientific American* (August 2007): 62. http://smc.neuralcorrelate.com/files/publications/martinez-conde_macknik_sciam07.pdf.

Susana Martinez-Conde, Stephen L. Macknik, Xoana G. Troncoso, and Thomas A. Dyarl. "Microsaccades Counteract Visual Fading during Fixation." *Neuron* 49 (January 19, 2006): 297–305. doi: 10.1016/j.neuron.2005.11.033. http://www.neuralcorrelate.com/smc/files/publications/martinez-conde_et_al_neuron06.pdf (subscription required).

Takatoshi Hikida, Satoshi Yawata, Takashi Yamaguchi, Teruko Danjo, Toshikuni Sasaoka, Yanyan Wang, and Shigetada Nakanishi. "Pathway-specific modulation of nucleus accumbens in reward and aversive behavior via selective transmitter receptors." *Proceedings of the National Academy of Sciences of the United States of America* 110, no. 1 (January 2, 2013): 342–47. doi: 10.1073/pnas.1220358110. http://www.ncbi.nlm.nih.gov/pmc/articles/PMC3538201.

Takemasa Yokoyama, Yasuki Noguchi, and Shinichi Kita. "Attentional shifts by gaze direction in voluntary orienting: Evidence from a microsaccade study." *Experimental Brain Research* 223, no. 2 (September 23, 2012): 291–300. doi: 10.1007/s00221-012-3260-z. http://www.ncbi.nlm.nih.gov/pmc/articles/PMC3475970.

Thomas H. Davenport. *Big Data @ Work*. Boston: Harvard Business Review Press, 2014.

Tim Higgins. "Apple Wants to Start Producing Cars as Soon as 2020." February 19, 2015. *Bloomberg*. http://www.bloomberg.com/news/articles/2015-02-19/

apple-said-to-be-targeting-car-production-as-soon-as-2020?hootPostID=9f3917f4905f086a568d81c5d794101a.

Timothy B. Lee. "Privacy groups seek investigation of Facebook's retail data sharing." September 27, 2012. Accessed February 25, 2015. *Ars Technica.* http://arstechnica.com/tech-policy/2012/09/privacy-groups-seek-investigation-of-facebooks-retail-data-sharing/.

Timothy Prickett Morgan. "A peek inside Apple's iCloud data center." June 9, 2011. *The Register.* http://www.theregister.co.uk/2011/06/09/apple_maiden_data_center/.

Tom Käckenhoff and Jean-Francois Rosnoblet. "REFILE-UPDATE 3—German pilot researched suicide, cockpit doors; second black box found." *Reuters,* April 2, 2015. http://www.reuters.com/article/2015/04/02/france-crash-recorder-update-3-tv-pixcor-idUSL6N0WZ31T20150402.

Tomio Geron. "Uber Confirms $258 Million from Google Ventures, TPG, Looks to On-Demand Future." *Forbes,* August 23, 2013. http://www.forbes.com/sites/tomiogeron/2013/08/23/uber-confirms-258-million-from-google-ventures-tpg-looks-to-on-demand-future/.

V. S. Chinmay. "Sensors on Google Glass." Accessed January 26, 2014. *The Code Artist Blog.* http://thecodeartist.blogspot.com/2013/05/sensors-on-google-glass.html.

Vance Packard. *The Hidden Persuaders.* New York: David McKay Company, 1957.

Verlyn Klinkenborg. "Trying to Measure the Amount of Information That Humans Create." Accessed February 23,

2015. *New York Times,* November 12, 2003. http://www.nytimes.com/2003/11/12/opinion/12WED4.html.

Victoria Woollaston. "New tracking software knows exactly where you'll be on a precise time and date YEARS into the future (even if you don't)." *Daily Mail.* http://www.dailymail.co.uk/sciencetech/article-2366566/Far-Out-software-knows-exactly-youll-precise-time-date-YEARS-future.html.

Vindu Goel. "Facebook Tinkers with Users' Emotions in News Feed Experiment, Stirring Outcry." *New York Times,* June 30, 2014. http://www.nytimes.com/2014/06/30/technology/facebook-tinkers-with-users-emotions-in-news-feed-experiment-stirring-outcry.html.

Yevgeniy Sverdlik. "UPDATED: Apple's Reno Data Center Expansion Marches On." August 8, 2014. *Data Center Knowledge.* http://www.datacenterknowledge.com/archives/2014/08/08/apple-data-center-expansion-in-reno-marches-on/.

———. "Apple Data Center Coming to Arizona." February 2, 2015. *Data Center Knowledge.* http://www.datacenterknowledge.com/archives/2015/02/02/apple-data-center-coming-to-arizona/.

Yuval Mor. "Emotions Analytics to Transform Human-Machine Interaction." *Huffington Post.* http://www.huffingtonpost.com/yuval-mor/emotions-analytics-to-tra_b_4240454.html.

Ziad M. Hafed. "Alteration of Visual Perception prior to Microsaccades." *Neuron* 77 (2013): 775–86. http://www.cnbc.cmu.edu/braingroup/papers/hafed_2013.pdf.

Ziad M. Hafed and James J. Clark. "Microsaccades as an overt measure of covert attention shifts." *Vision Research* 42 (2002): 2,533–2,545. http://www.cim.mcgill.ca/~clark/vmrl/web-content/papers/jjclark_vr_22_2002.pdf or http://www.physiol-active-vision.uni-tuebingen.de/paper/hafed_vis_res_reprint2002.pdf.

Listed by Government Agency

Department of Homeland Security

"Behavioral biometrics to detect terrorists entering U.S." September 28, 2010. *Homeland Security News Wire*. http://www.homelandsecuritynewswire.com/behavioral-biometrics-detect-terrorists-entering-us.

"Future Attribute Screening Technology (FAST) Demonstration Laboratory—HSARPA BAA07-03A." September 21, 2007. Accessed April 19, 2015. *FedBizOpps.Gov*. https://www.fbo.gov/index?s=opportunity&mode=form&tab=core&id=e7535b0bdd3e1609e1d840fdbd26c114.

United States District Court, Northern California, San Jose Division (litigation)

Google, Inc., Gmail litigation, case no. 13-MD-02430-LHK, "Consented to Automated Scanning Pursuant to § 2511(2)(d)" (pages 20–24), United States District Court, Northern District of California, San Jose Division, October 9, 2013, signed by Kathleen M. Sullivan, attorney for the defendant, accessed December 15, 2014.

Google, Inc., Gmail litigation, case no. 13-MD-02430-LHK, Lucy H. Koh, United States district judge, March 18,

2014, http://www.consumerwatchdog.org/resources/googlemotion061313.pdf.

Federal Trade Commission

United States v. Snapchat, Inc. https://www.ftc.gov/system/files/documents/cases/140508snapchatcmpt.pdf.

United States v. Facebook. https://www.ftc.gov/news-events/press-releases/2011/11/facebook-settles-ftc-charges-it-deceived-consumers-failing-keep.

United States v. Google, Inc. https://www.ftc.gov/enforcement/cases-proceedings/102-3136/google-inc-matter.

Office of the Under Secretary of Defense for Acquisition, Technology, and Logistics

"MilCon Status Report—August 2014—Under Secretary of Defense for AT&L," September 17, 2014, accessed February 22, 2015, http://www.acq.osd.mil/ie/fim/library/milcon/MILCON_EOM-AUG_Report_2014-09-17.xlsx, line 1,607.

Securities and Exchange Commission

Amendment 4 to Form S-1, "Registration Statement," Google, page 8. Securities and Exchange Commission. http://www.sec.gov/Archives/edgar/data/1288776/000119312504124025/ds1a.htm.

US Attorney General

Gonzales v. Google, Inc., filing 12. *Justia*. http://docs.justia.com/cases/federal/district-courts/california/candce/5:2006mc80006/175448/12/.

Listed by Patent (assignee, inventor, or title, whichever is likely more relevant to the reader)

Apple, Inc., US patent application no. 20130144789, "Method and System for Managing Credits via a Mobile Device," http://appft.uspto.gov/netacgi/nph-Parser?Sect1=PTO2&Sect2=HITOFF&p=1&u=%2Fnetahtml%2FPTO%2Fsearch-adv.html&r=1&f=G&l=50&d=PG01&S1=%28%22Managing+Credits%22.TTL.%29&OS=TTL/%22+Managing+Credits%22&RS=TTL/%22+Managing+Credits%22.

Apple, Inc., US patent application no. 13/556023, "Inferring User Mood Based on User and Group Characteristic Data," *US Patent and Trademark Office,* http://appft.uspto.gov/netacgi/nph-Parser?Sect1=PTO1&Sect2=HITOFF&d=PG01&p=1&u=%2Fnetahtml%2FPTO%2Fsrchnum.html&r=1&f=G&l=50&s1=%2220140025620%22.PGNR.&OS=DN/20140025620&RS=DN/20140025620.

Apple, Inc., US patent 8,937,591, "Systems and methods for counteracting a perceptual fading of a movable indicator," filed April 6, 2012, and issued January 20, 2015, accessed April 9, 2015, http://patft.uspto.gov/netacgi/nph-Parser?Sect1=PTO2&Sect2=HITOFF&u=%2Fnetahtml%2FPTO%2Fsearch-adv.htm&r=2&p=1&f=G&l=50&d=PTXT&S1=%28345%2F157.CCLS.+AND+20150120.PD.%29&OS=ccl/345/157+and+isd/1/20/2015&RS=%28CCL/345/157+AND+ISD/20150120%29, claims 10 and 27.

Google (Raffle, Hayes Solos, Adrian Wong, and Ryan Geiss), US patent 8,235,529, "Unlocking a screen with eye tracking information," http://pdfpiw.uspto.gov/.piw?PageNum=0&docid=08235529&IDKey=DB61E66D7179%0D%0A&HomeUrl=http%3A%2F%2Fpatft.uspto.gov%2Fnetahtml%2FPTO%2Fpatimg.htm.

Office of Naval Research, grant no. F49620-97-1-0353, and Air Force Office of Scientific Research, grant no. N00014-93-1-0525; from Sandra P. Marshall, US patent 6,090,051, "Statement of Government Rights: Method and apparatus for eye tracking and monitoring pupil dilation to evaluate cognitive activity," filed March 3, 1999 and issued July 18, 2000, accessed February 25, 2015. http://patft.uspto.gov/netacgi/nph-Parser?Sect1=PTO1&Sect2=HITOFF&d=PALL&p=1&u=%2Fnetahtml%2FPTO%2Fsrchnum.htm&r=1&f=G&l=50&s1=6,090,051.PN.&OS=PN/6,090,051&RS=PN/6,090,051.

Nielsen Company (Anantha Pradeep, Robert T. Knight, and Ramachandran Gurumoorthy), US patent 8,494,905, "Audience response analysis using simultaneous electroencephalography (EEG) and functional magnetic resonance imaging (fMRI)", http://pdfpiw.uspto.gov/.piw?PageNum=0&docid=08494905&IDKey=AFD0F456F6C5%0D%0A&HomeUrl=http%3A%2F%2Fpatft.uspto.gov%2Fnetahtml%2FPTO%2Fpatimg.htm.

Sandra P. Marshall, US patent 6,090,051, "Method and apparatus for eye tracking and monitoring pupil dilation to evaluate cognitive activity," http://patft.uspto.gov/netacgi/nph-Parser?Sect1=PTO1&Sect2=HITOFF&d=PALL&p=1&u=%2Fnetahtml%2FPTO%2Fsrchnum.htm&r=1&f=G&l=50&s1=6,090,051.PN.&OS=PN/6,090,051&RS=PN/6,090,051.

Theodoros G. Paraskevakos, US patent 3,842,208, "Sensor Monitoring Device," http://patft.uspto.gov/netacgi/nph-Parser?Sect1=PTO1&Sect2=HITOFF&d=PALL&p=1&u=%2Fnetahtml%2FPTO%2Fsrchnum.htm&r=1&f=G&l=50&s1=3,842,208.PN.&OS=PN/3,842,208&RS=PN/3,842,208.

Theodoros G. Paraskevakos, US patent 4,241,237, "Apparatus and method for remote sensor monitoring, metering and control," filed January 26, 1979, and issued December 23, 1980, http://patft.uspto.gov/netacgi/nph-Parser?Sect1=PTO2&Sect2=HITOFF&p=1&u=%2Fnetahtml%2FPTO%2Fsearch-bool.html&r=1&f=G&l=50&co1=AND&d=PTXT&s1=%22Paraskevakos%3B+Theodoros+G%22.INNM.&OS=IN/.

Listed by Subject or Topic

"Augmented Reality for Glass." *Augmented Reality for Glass.* http://arforglass.org/.

Brain Waves

"Alpha Brain Waves Background Information." *Brain Waves Blog.* http://www.brainwavesblog.com/tag/alpha-waves/.

"Alpha Brain Waves Background Information." *Brain Waves Blog.* http://www.brainwavesblog.com/alpha-brain-waves-information/.

"Gamma Brain Waves Information." *Brain Waves Blog.* http://www.brainwavesblog.com/gamma-brain-waves-information/.

"What Are Gamma Brain Waves?" *Brain Waves Blog.* http://www.brainwavesblog.com/gamma-brain-waves-information/.

"Collecting private information, Uses and abuses, A computer-security expert weighs up the costs and benefits of collecting masses of personal data." *The Economist.* http://www.economist.com/news/books-and-arts/21647595-computer-security-expert-weighs-up-costs-and-benefits-collecting-masses.

"Smartphone Apps Covertly Report Your Location Data" Uploaded on April 3, 2015, Cyber Security Intelligence, accessed June 12, 2015 http://www.cybersecurityintelligence.com/blog/smartphone-apps-covertly-report-your-location-data--189.html.

"EU says Google hurt consumers and competitors in Internet search case." *Reuters,* April 15, 2015. http://www.reuters.com/article/2015/04/15/us-google-eu-idUSKBN0N610E20150415.

"Facebook Data Center FAQ." *Data Center Knowledge.* http://www.datacenterknowledge.com/the-facebook-data-center-faq/.

"Google Glass Can Read Your Mind." *Kirkus Reviews.* https://www.kirkusreviews.com/book-reviews/jack-m-wedam/google-glass-can-read-your-mind/.

"Google Glass makes doctors better surgeons: A Stanford study." September 29, 2014. *Surgery Academy.* http://www.surgeryacade.my/2014/09/29/google-glass-makes-doctors-better-surgeons-a-stanford-study/.

"Google Will Fight Government Over Access to Your Emails." *Reuters,* January 28, 2013. http://www.reuters.com/article/2013/01/28/us-usa-privacy-google-idUSBRE90R17120130128.

"iPhone 6 Eye Tracking." *iPhone News*. http://iphoneipadipod.com/apple/iphone-6-eye-tracking.html.

"MilCon Status Report—August 2014—Under Secretary of Defense for AT&L." September 17, 2014. Office of the Under Secretary of Defense for Acquisition, Technology, and Logistics. http://www.acq.osd.mil/ie/fim/library/milcon/MILCON_EOM-AUG_Report_2014-09-17.xlsx.

"Paul Lazarsfeld Guest Professorship." University of Vienna. http://methods.univie.ac.at/paul-lazarsfeld-professorship/.

"Sabina, a Robot Domestic Learns When You Show Her." *Science 2.0*. http://www.science20.com/news_articles/sabina_a_robot_domestic_learns_when_you_show_her-154512.

"Software: Cognitive Workload." Accessed February 25, 2015. *EyeTracking, Inc.* http://www.eyetracking.com/Software/Cognitive-Workload.

"Targeted Online Advertising: Data Worth Their Weight in Gold 2009," in *30th Activity Report*. Commission Nationale de l'informatique et des Libertés, http://www.cnil.fr/fileadmin/documents/en/CNIL-30e_rapport_2009-EN.pdf.

"Tobii Eye Tracking: See it from her point of view." *Acuity ETS*. http://acuity-ets.com/downloads/tobii_mr_brochure_web.pdf.

"Understanding TCP/IP addressing and subnetting basics." *Microsoft*. http://support.microsoft.com/kb/164015.

"Unfriending Cash." *The Economist*. http://www.economist.com/news/finance-and-economics/21646802-facebook-enters-booming-market-mobile-payments-unfriending-cash.

"What the Eyes Reveal: 10 Messages My Pupils Are Sending You." December 2011. *PsyBlog*. http://www.spring.org.uk/2011/12/what-the-eyes-reveal-10-messages-my-pupils-are-sending-you.php.

Listed by Website

Apple. http://www.apple.com/watch/technology/#familiar.

Autodesk 123D. http://www.123dapp.com/catch.

Beyond Verbal. http://www.beyondverbal.com/.

Data Center Knowledge. "Facebook Data Center FAQ." http://www.datacenterknowledge.com/the-facebook-data-center-faq/.

Der Spiegel. http://www.spiegel.de/international/.

Dictionary.com (s.v. "neuromarketing").

Ello. https://ello.co/wtf/post/about.

Emotiv. http://www.emotiv.com/.

Free Medical Dictionary (s.v. "saccades"; s.v. "psychophysiological").

Future of Life Institute. "Research Priorities for Robust and Beneficial Artificial Intelligence: An Open Letter." http://futureoflife.org/misc/open_letter.Google Data Centers

Ginger.io. https://ginger.io/.

Google.

"How does Google protect my privacy and keep my information secure?" http://www.google.com/policies/faq/.

"AI research at Google, Artificial Brains." http://www.artificialbrains.com/google.

Data Centers

Locations: http://www.google.com/about/datacenters/inside/locations/index.html.

Mayes County, Oklahoma: http://www.google.com/about/datacenters/inside/locations/mayes-county/.

Berkeley County, South Carolina: http://www.google.com/about/datacenters/inside/locations/berkeley-county/.

Council Bluffs, Iowa: http://www.google.com/about/datacenters/inside/locations/council_bluffs/.

Health and Human Services. "Massachusetts provider settles HIPAA case for $1.5 million." *HHS.gov.* http://www.hhs.gov/ocr/privacy/hipaa/enforcement/examples/meei-agreement.html.

IT Law wiki. "Anonymous-entity resolution." http://itlaw.wikia.com/wiki/Anonymous-entity_resolution.

Jean Kilbourne. http://www.jeankilbourne.com/bio/.

Mandala Consulting. http://www.mandalaconsulting.co.za/Documents/Reviews/Review%201.pdf.

Merriam-Webster Dictionary (s.v. "algorithm"; s.v. "autonomous"; s.v. "nucleus accumbens"). http://www.merriam-webster.com/dictionary/.

Neuro-Insight. http://www.neuro-insight.com/.

Nielsen. http://www.nielsen.com/us/en/solutions/measurement.html.

Phys.Org. "Samsung introduces EYECAN+, next-generation mouse for people with disabilities." http://phys.org/news/2014-11-samsung-eyecan-next-generation-mouse-people.html#jCp.

Primary Objects. "Using Artificial Intelligence to Write Self-Modifying/Improving Programs." http://www.primaryobjects.com/CMS/Article149.

Puzzlebox.io. http://puzzlebox.io/orbit/.

PredictiveAnalytics. http://predictiveanalytics.org/.

Predictive Analytics World. http://www.predictiveanalyticsworld.com/.

Sash Company. https://www.thesashcompany.com/missamerica.php.

Sensory Logic. http://www.sensorylogic.com/.

SRI International. http://www.sri.com/sites/default/files/brochures/sri-overview.pdf p 2.

The Brain, Top to Bottom. http://thebrain.mcgill.ca/flash/a/a_03/a_03_cr/a_03_cr_par/a_03_cr_par.html.

Wikipedia.

"Affectiva." *Wikipedia.* Last updated January 15, 2015. Accessed June 2, 2015. http://en.wikipedia.org/wiki/Affectiva.

"The Amygdaloids." *Wikipedia*. Last updated January 27, 2015. Accessed June 15, 2015. https://en.wikipedia.org/wiki/The_Amygdaloids.

"Apple Watch." *Wikipedia*. Last updated June 15, 2015. Accessed June 15, 2015. https://en.wikipedia.org/wiki/Apple_Watch.

"Benjamin Libet." *Wikipedia*. Last updated May 2, 2015. Accessed June 15, 2015. https://en.wikipedia.org/wiki/Benjamin_Libet.

"Cryptographic hash function," *Wikipedia*. last updated June 3, 2015. Accessed February 25, 2015. http://en.wikipedia.org/wiki/Cryptographic_hash_function.

"Digital identity." *Wikipedia*. Last updated June 12, 2015. Accessed June 15, 2015. https://en.wikipedia.org/wiki/Digital_identity.

"Electroencephalography." *Wikipedia*. Last updated June 13, 2015. Accessed June 15, 2015. https://en.wikipedia.org/wiki/Electroencephalography.

"Facial Action Coding System." *Wikipedia*. Last updated May 5, 2015. Accessed June 15, 2015. https://en.wikipedia.org/wiki/Facial_Action_Coding_System.

"Fear." *Wikipedia*. Last updated 9 June 2015. Accessed June 15, 2015. https://en.wikipedia.org/wiki/Fear.

"Functional magnetic resonance imaging." *Wikipedia*. Last updated 30 May 2015. Accessed June 15, 2015. https://en.wikipedia.org/wiki/Functional_magnetic_resonance_imaging.

"Future Attribute Screening Technology." *Wikipedia*. Last updated September 18, 2014. Accessed June 15, 2015. https://en.wikipedia.org/wiki/Future_Attribute_Screening_Technology.

"Google Glass." *Wikipedia*. Last updated 1 June 2015. Accessed June 15, 2015. https://en.wikipedia.org/wiki/Google_Glass.

"Hash function." *Wikipedia*. Last updated June 11, 2015. Accessed June 15, 2015. https://en.wikipedia.org/wiki/Hash_function.

"HTTP cookie." *Wikipedia*. Last updated June 15, 2015. Accessed June 15, 2015. https://en.wikipedia.org/wiki/HTTP_cookie.

"Record linkage" redirected from "Identity resolution." *Wikipedia*. Last updated June 4, 2015 Accessed June 15, 2015. https://en.wikipedia.org/wiki/Record_linkage#Identity_resolution.

"IP address." *Wikipedia*. Last updated 23 May 23, 2015, Accessed June 15, 2015. https://en.wikipedia.org/wiki/IP_address.

"Joseph E. LeDoux." *Wikipedia*. Last updated June 6, 2015. Accessed June 15, 2015. https://en.wikipedia.org/wiki/Joseph_E._LeDoux.

"Mathematical sociology." *Wikipedia*. Last updated May 26, 2015. Accessed June 15, 2015. https://en.wikipedia.org/wiki/Mathematical_sociology.

"Military-industrial complex." *Wikipedia*. Last updated June 6, 2015. Accessed June

15, 2015. https://en.wikipedia.org/wiki/ Military%E2%80%93industrial_complex.

"Marlboro (cigarette)." *Wikipedia*. Last updated June 12, 2015. Accessed June 15, 2015. https://en.wikipedia.org/wiki/Marlboro_%28cigarette%29.

"Marlboro Man." *Wikipedia*. Last updated 18 May 18, 2015. Accessed June 16, 2015. https://en.wikipedia.org/wiki/Marlboro_Man.

"Minority Report (film)." *Wikipedia*. Last updated June 11, 2015. Accessed June 15, 2015. https://en.wikipedia.org/wiki/Minority_Report_%28film%29.

"Neuromarketing." *Wikipedia*. Last updated April 13, 2015. Accessed June 15, 2015. https://en.wikipedia.org/wiki/Neuromarketing.

"No Lie MRI." *Wikipedia*. Last updated 9 July 2014. Accessed June 15, 2015. https://en.wikipedia.org/wiki/No_Lie_MRI.

"Nucleus accumbens." *Wikipedia*. Last updated April 17, 2015. Accessed June 15, 2015. https://en.wikipedia.org/wiki/Nucleus_accumbens.

"Paul Ekman." *Wikipedia*. Last updated May 31, 2015. Accessed June 15, 2015. https://en.wikipedia.org/wiki/Paul_Ekman.

"Paul Lazarsfeld." *Wikipedia*. Last updated May 4, 2015. Accessed June 15, 2015. https://en.wikipedia.org/wiki/Paul_Lazarsfeld.

"Personally identifiable information." *Wikipedia.* Last updated June 8, 2015. Accessed June 15, 2015. https://en.wikipedia.org/wiki/Personally_identifiable_information.

"Saccade." *Wikipedia.* Last updated June 10, 2015. Accessed June 15, 2015. https://en.wikipedia.org/wiki/Saccade.

"Sexual network." *Wikipedia.* Last updated April 2, 2015. Accessed June 15, 2015. https://en.wikipedia.org/wiki/Sexual_network.

"Silvan Tomkins." *Wikipedia.* Last updated November 9, 2014. Accessed June 15, 2015. https://en.wikipedia.org/wiki/Silvan_Tomkins.

"Smartphone." *Wikipedia.* Last updated June 10, 2015. Accessed June 15, 2015. https://en.wikipedia.org/wiki/Smartphone.

"Social network." *Wikipedia.* Last updated June 9, 2015. Accessed June 15, 2015. https://en.wikipedia.org/wiki/Social_network.

"Social network analysis." *Wikipedia.* Last updated 30 May 30, 2015. Accessed June 15, 2015. https://en.wikipedia.org/wiki/Social_network_analysis.

"Steady state topography." *Wikipedia.* Last updated May 26, 2015. Accessed June 15, 2015. https://en.wikipedia.org/wiki/Steady_state_topography.

"Theodore Paraskevakos." *Wikipedia,* last updated 6 September 6, 2014 accessed April 20, 2015. https://en.wikipedia.org/wiki/Theodore_Paraskevakos.

"Utah Data Center." *Wikipedia.* **Last updated 13 June 2015. Accessed June 15, 2015.** https://en.wikipedia.org/wiki/Utah_Data_Center.

Notes

Author's Preface

1. "Department of Homeland Security, Future Attribute Screening Technology (FAST) Demonstration Laboratory—HSARPA BAA07-03A," *FedBizOpps,* September 21, 2007, accessed April 19, 2015, https://www.fbo.gov/index?s=opportunity&mode=form&tab=core&id=e7535b0bdd3e1609e1d840fdbd26c114.

    "Future Attribute Screening Technology," *Wikipedia,* last updated September 18, 2014, Accessed June 15, 2015, http://en.wikipedia.org/wiki/Future_Attribute_Screening_Technology.

2. "Behavioral biometrics to detect terrorists entering U.S.," *Homeland Security News Wire,* September 28, 2010, accessed April 19, 2015, http://www.homelandsecuritynewswire.com/behavioral-biometrics-detect-terrorists-entering-us.

3. Guy Grimland, "Israel startup uses behavioral science to identify terrorists," *Haaretz,* May 9, 2008, accessed April 19, 2015, http://www.haaretz.com/print-edition/business/israel-startup-uses-behavioral-science-to-identify-terrorists-1.245470.

4. Phillip Bonacich and Philp Lu, *Introduction to Mathematical Sociology* (Princeton: Princeton University Press, 2012), 1.

5. Jack M. Wedam, US patent 4,654,647, "Finger actuated electronic

control apparatus," filed September 24, 1984, and issued March 31, 1987, *United States Patent and Trademark Office,* accessed May 20, 2014, http://patft.uspto.gov/netacgi/nph-Parser?Sect1 =PTO1&Sect2=HITOFF&d=PALL&p=1&u=%2Fnetahtml%2FP TO%2Fsrchnum.htm&r=1&f=G&l=50&s1=4,654,647.PN.&OS= PN/4,654,647&RS=PN/4,654,647.

Introduction

1. The Flesch–Kincaid reading level for the main body of text of *Cunningly Smart Phones* (minus the glossary, bibliography, references, and notes) is 11.9. The Flesch level of reading ease is 44.3. For a comparison, Florida requires insurance policies to have a Flesch reading-ease score of 45 or higher (http://law.onecle.com/florida/insurance/627.4145.html). Therefore, if you have earned a high school diploma (or the equivalent) and you can read a Florida insurance policy, then you should be able to read this book.

A Cunningly Smart Spy

1. Elizabeth Dwoskin, "What Secrets Your Phone Is Sharing about You," *Wall Street Journal,* January 13, 2014, accessed March 4, 2015, http://online.wsj.com/news/articles/SB10001424052702303453004579290632128929194?mod=trending now 1.

2. Bruce Schneier, *Data and Goliath: The Hidden Battle to Collect Your Data and Control You* (New York: W. W. Norton, 2015), 108–18.

3. Julia Angwin, "The Web's New Gold Mine: Your Secrets," *Wall Street Journal,* July 30, 2010, accessed April 13, 2015, http://online.wsj.com/article/SB10

001424052748703940904575395073512989404.html?mod=WSJ_WhatTheyKnow2010_WhatsNews_4_2_Left.

4   Anantha Pradeep, Robert T. Knight, and Ramachandran Gurumoorthy, US patent 8,494,905, filed June 6, 2008, and issued July 23, 2013, column 3, lines 47–59: "accurately reflect a subject's actual thoughts. ... Consequently, the techniques and mechanisms of the present invention intelligently blend multiple modes such as EEG and fMRI to more accurately assess effectiveness of stimulus materials."

5   Michael Froimowitz Greenzeiger, Ravindra, Phulari, and Mehul K. Sanghavi, "Inferring User Mood Based on User and Group Characteristic Data," United States patent application no. 13/556023, filed July 23, 2012, *United States Patent and Trademark Office,* accessed January 17, 2015, http://appft.uspto.gov/netacgi/nph-Parser?Sect1=PTO1&Sect2=HITOFF&d=PG01&p=1&u=%2Fnetahtml%2FPTO%2Fsrchnum.html&r=1&f=G&l=50&s1=%2220140025620%22.PGNR.&OS=DN/20140025620&RS=DN/20140025620.

6   Schneier, *Data and Goliath,* 2.

7   Victoria Woollaston, "New tracking software knows exactly where you'll be on a precise time and date YEARS into the future (even if you don't)," *Daily Mail,* July 17, 2013, accessed March 31, 2015, http://www.dailymail.co.uk/sciencetech/article-2366566/Far-Out-software-knows-exactly-youll-precise-time-date-YEARS-future.html.

8   Kashmir Hill, "How Target Figured Out a Teen Girl Was Pregnant before Her Father Did," *Forbes,* February 16, 2012, accessed April 8, 2015, http://www.forbes.com/sites/kashmirhill/2012/02/16/how-target-figured-out-a-teen-girl-was-pregnant-before-her-father-did/.

Charles Duhigg, "How Companies Learn Your Secrets," *New York Times,* February 19, 2012, accessed February 12, 2015, http://www.nytimes.com/2012/02/19/magazine/shopping-habits.html?pagewanted=all&_r=0.

9   Nick Saint, "Google CEO: 'We Know Where You Are. We Know Where You've Been. We Can More Or Less Know What You're Thinking About,'" October 4, 2010, *Business Insider,* October 2010, accessed March 25, 2015, http://www.businessinsider.com/eric-schmidt-we-know-where-you-are-we-know-where-youve-been-we-can-more-or-less-know-what-youre-thinking-about-2010-10.

10   Martin Lindstrom, *Buyology* (New York: Crown Business, 2010), 14–16, 82–83, 196–7.

11   "Collecting private information: Uses and abuses: A computer-security expert weighs up the costs and benefits of collecting masses of personal data," *The Economist,* April 4, 2015, accessed April 19, 2015, http://www.economist.com/news/books-and-arts/21647595-computer-security-expert-weighs-up-costs-and-benefits-collecting-masses.

12   Emily Steel, "Companies scramble for consumer data," *Financial Times,* June 12, 2013, accessed March 3, 2015, http://www.ft.com/intl/cms/s/0/f0b6edc0-d342-11e2-b3ff-00144feab7de.html#axzz2W71Wg6oo (subscription required).

13   Angwin, "The Web's New Gold Mine."

14   Brody Mullins, Rolfe Winkler, and Brent Kendall, "Inside the U.S. Antitrust Probe of Google," *Wall Street Journal,* March 19, 2015, accessed March 20, 2015, http://www.wsj.com/articles/inside-the-u-s-antitrust-probe-of-google-1426793274.

15 "Theodore Paraskevakos," *Wikipedia,* last updated 6 September 6, 2014 accessed April 20, 2015. https://en.wikipedia.org/wiki/Theodore_Paraskevakos.

Of the more than fifty patents worldwide, the two patents relevant to the development of the smartphone are the following:

Theodoros G. Paraskevakos, US patent 3,842,208, "Sensor Monitoring Device," filed July 11, 1972, and issued October 15, 1974, *United States Patent and Trademark Office,* accessed April 6, 2015, http://patft.uspto.gov/netacgi/nph-Parser?Sect1=PTO1&Sect2=HITOFF&d=PALL&p=1&u=%2Fnetahtml%2FPTO%2Fsrchnum.htm&r=1&f=G&l=50&s1=3,842,208.PN.&OS=PN/3,842,208&RS=PN/3,842,208.

Theodoros G. Paraskevakos, US patent 4,241,237, "Apparatus and method for remote sensor monitoring, metering and control," filed January 26, 1979, and issued December 23, 1980, *United States Patent and Trademark Office,* http://patft.uspto.gov/netacgi/nph-Parser?Sect1=PTO2&Sect2=HITOFF&p=1&u=%2Fnetahtml%2FPTO%2Fsearch-bool.html&r=1&f=G&l=50&col=AND&d=PTXT&s1=%22Paraskevakos%3B+Theodoros+G%22.INNM.&OS=IN/.

16 "Smartphone," *Wikipedia,* last updated June 10, 2015, accessed June 15, 2015, http://en.wikipedia.org/wiki/Smartphone#Early_years.

17 "Bombe," *Crypto Museum,* February 7, 2014, accessed March 21, 2015, http://www.cryptomuseum.com/crypto/bombe/.

18 Many other smart devices are being commercialized, including smart watches, smart glasses (Samsung's version can be seen here: http://www.pcmag.com/article2/0,2817,2429969,00.asp), and smart bicycles (http://www.reuters.com/video/2015/02/26/

vibrating-bicycle-senses-traffic?videoId=363309219&videoChannel=6). However, the number of units that were sold and are currently used by smartphones is much larger than the number of all other smart devices sold and in use as of 2015.

19   "Smartphone security: The spy in your pocket," *The Economist,* February 28, 2015, accessed March 5, 2015, http://www.economist.com/news/briefing/21645130-watch-out-hackersand-spooks-spy-your-pocket.

20   Andrew Griffin, "iOS 9: iPhone will now track sexual activity," *The Independent,* 09 June 2015, accessed June 11, 2015, http://www.independent.co.uk/life-style/gadgets-and-tech/news/ios-9-iphone-will-now-track-sexual-activity-10307408.html .

21   Schneier, *Data and Goliath,* 16.

22   Thomas H. Davenport, *Big Data @ Work* (Boston: Harvard Business Review Press, 2014), 27.

23   Mary Madden, "Public Perceptions of Privacy and Security in the Post-Snowden Era," *Pew Research Center,* November 12, 2014, accessed January 28, 2015, http://pewrsr.ch/1wlEbYR.

- Ninety-one percent of adults in the survey agree or strongly agree that consumers have lost control over how personal information is collected and used by companies.

- Eighty-eight percent of adults agree or strongly agree that it would be very difficult to remove inaccurate information about them online.

- Eighty percent of those who use social networking sites say they are concerned about third parties like advertisers and businesses accessing the data they share on these sites.

- Seventy percent of social networking site users say that they are at least somewhat concerned about the government accessing some of the information they share on social networking sites without their knowledge.

24 "Google Glass Can Read Your Mind," *Kirkus Reviews,* accessed April 12, 2015, https://www.kirkusreviews.com/book-reviews/jack-m-wedam/google-glass-can-read-your-mind/.

25 Ibid.

26 Madden, "Public Perceptions."

- Sixty-one percent of adults disagree or strongly disagree with the statement, "I appreciate that online services are more efficient because of the increased access they have to my personal data."

- At the same time, 55 percent agree or strongly agree with the statement, "I am willing to share some information about myself with companies in order to use online services for free."

27 "Executive Summary: Data Brokers: A Call for Transparency and Accountability," *Federal Trade Commission,* May 27, 2014, accessed May 27, 2014, http://www.ftc.gov/system/files/documents/reports/data-brokers-call-transparency-accountability-report-federal-trade-commission-may-2014/140527databrokerreport.pdf.

28 Dyan Machan, "The New Information Goldmine," *Wall Street Journal,* August 19, 2009, accessed March 20, 2015, http://online.wsj.com/article/SB125071202052143965.html.

Angwin, "The Web's New Gold Mine."

David Finn, "Health Information = A Hacker's Gold Mine," *Symantec,* June 18, 2012, accessed March 20, 2015, http://www.symantec.com/connect/blogs/health-information-hacker-s-gold-mine.

Heather Green, "The Information Gold Mine," *Business Week,* July 26, 1999, accessed March 8, 2015, http://www.businessweek.com/1999/99_30/b3639030.htm.

"30th Activity Report: Targeted Online Advertising: Data Worth Their Weight in Gold 2009," *Commission Nationale de l'informatique et des Libertés,* accessed March 20, 2015, http://www.cnil.fr/fileadmin/documents/en/CNIL-30e_rapport_2009-EN.pdf, 14.

"ENISA: Smartphones a 'goldmine of sensitive and personal information': EU Cyber-security research from ENISA highlights the key risks from smartphone usage," *Dennis Publishing,* accessed May 5, 2013, http://www.knowyourmobile.com/products/11341/enisa-smartphones-goldmine-sensitive-and-personal-information.

[29] Green, "The Information Gold Mine."

[30] "EU says Google hurt consumers and competitors in Internet search case," *Reuters,* April 15, 2015, accessed April 15, 2015, http://www.reuters.com/article/2015/04/15/us-google-eu-idUSKBN0N610E20150415.

[31] Matt Rosoff, "Is Google a Monopoly? 'We're in that Area,' Admits Schmidt," *Business Insider,* accessed April 14, 2015, http://articles.businessinsider.com/2011-09-21/tech/30183638_1_monopoly-web-browser-market-microsoft.

[32] "Internet Sheriff," *Harvard Magazine,* accessed April 4, 2013, http://harvardmagazine.com/2010/09/internet-sheriff.

33 "Protecting privacy online; The price of reputation; Is the market for protected personal information about to take off?" *The Economist,* accessed January 17, 2015, http://www.economist.com/news/business/21572240-market-protected-personal-information-about-take-price-reputation.

"Natasha Singer, A Vault for Taking Charge of Your Online Life," *New York Times,* December 8, 2012, accessed April 8, 2015, http://www.nytimes.com/2012/12/09/business/company-envisions-vaults-for-personal-data.html?pagewanted=all.

34 Among a mere five corporations in 2013, $77,367,000,000 in revenues were earned by selling users' personal information (Google, $58.807 billion; Facebook, $7.782 billion; Baidu, $5.277 billion; Alliance Data Systems Corporation, $4.319 billion; Acxiom, $1.093 billion). The $77 billion estimate excludes other social media corporations, data brokers, and data analyzers of transaction-rich data that sell your information to marketers and advertisers. Including eight more data brokers (Corelogic, Datalogix, eBureau, ID Analytics, Intelius, PeekYou, Rapleaf, and Recorded Future) would result in higher revenues. Therefore, the figure mentioned is a low estimate.

Assuming that everyone in the world is equal and that $77,367,000,000 (in revenue) is divided by 7,095,217,980 (the number of people in the world) (*CIA World Factbook,* 2013), this results in $10.90 per year for every person on earth. This amount represents a crude estimate of the market value of personal information per person on a global basis. This estimate can be refined by adjusting for differences in purchasing power. The gross domestic product (GDP) in price purchasing parity (PPP) in 2013 was $13,100 for everyone on earth (*CIA Factbook,* 2013, https://www.cia.gov/library/publications/the-world-factbook/geos/xx.html). Scaling this up for the United States, where GDP in PPP was $52,800 in 2013, results in $43.95 per person per year for the value of personal information ([52,800 ÷ 13,100] × 10.90). Forty-three dollars and ninety cents per year multiplied by twenty-five years is over a thousand dollars ($1,098.74).

The amount varies with your disposable income. The value of personal information is likely to be much higher for a thirty-year-old earning seventy thousand dollars per year than a two-year-old or an eighty-year-old, since the thirty-year-old is more likely to be making decisions about spending money or voting than are the young and the old. Furthermore, the value of personal information in this estimate would be even higher if the revenues from all the corporations that sell personal information were included instead of only five corporations (Google, Facebook, Baidu, Alliance Data Systems, and Acxiom).

Year 2013 is used as the reference year, since the 2014 world PPP estimate was not available from the *CIA Factbook* at the time of publication. However, the revenue of the five corporations listed above increased from $77 billion to $92 billion. Therefore, per-person estimates above are low estimates and are significantly higher for the year 2014.

[35] "John Wanamaker," *BrainyQuote*, accessed May 17, 2015, https://www.brainyquote.com/quotes/authors/j/john_wanamaker.html.

[36] Amber Haq, "This Is Your Brain on Advertising," *Bloomberg Business*, October 8, 2007, accessed March 7, 2015, http://www.bloomberg.com/bw/stories/2007-10-08/this-is-your-brain-on-advertisingbusinessweek-business-news-stock-market-and-financial-advice.

Mary Godwin

1. Mary Wollstonecraft Shelley, *Frankenstein; or, The Modern Prometheus* (Garden City, NY: Halcyon House, 2010), 171.

2. Shelley, *Frankenstein*, 177.

3. Wi-Fi, fiber optics, cell phone towers, etc.

4. Electronic memory, "cloud memory," etc.

5. Computer programs, algorithms, smartphone apps, etc.

6. Shelley, *Frankenstein*, 177.

7. Ibid. 171.

8. Ibid, 177.

9. "Paul Ekman," *Wikipedia*, last updated May 31, 2015, accessed June 15, 2015, http://en.wikipedia.org/wiki/Paul_Ekman.

10. "Facial Action Coding System," *Wikipedia*, last updated May 5, 2015, accessed June 15, 2015, http://en.wikipedia.org/wiki/Facial_Action_Coding_System.

11. Elizabeth Dwoskin and Evelyn M. Rusli, "The Technology that Unmasks Your Hidden Emotions," *Wall Street Journal*, January 28, 2015, accessed January 31, 2015, http://www.wsj.com/articles/startups-see-your-face-unmask-your-emotions-1422472398?mod=LS1.

What's Emotion Got to Do with It?

[1] "Joseph E. LeDoux," *Wikipedia,* last updated June 6, 2015, accessed June 15, 2015, http://en.wikipedia.org/wiki/Joseph_E._LeDoux.

[2] "The Amygdaloids," *Wikipedia,* last updated January 27, 2015, accessed June 15, 2015, http://en.wikipedia.org/wiki/The_Amygdaloids.

[3] Hamish Pringle, "Why Emotional Messages Beat Rational Ones," *Advertising Age,* March 2, 2009, accessed March 17, 2015, http://adage.com/article/cmo-strategy/emotional-messages-beat-rational/134920/2015.

Hamish Pringle and Peter Field, *Brand Immortality* (Philadelphia: Kogan Page, 2008), 176.

[4] Jonah Lehrer, *How We Decide* (New York: Houghton Miffin Harcourt, 2009), 18.

[5] Douglas van Praet, *Unconscious Branding: How Neuroscience Can Empower (and Inspire) Marketing* (New York: Palgrave Macmillan, 2012), 61.

[6] Joseph LeDoux, *The Emotional Brain: The Mysterious Understanding of Emotions* (New York: Simon and Schuster, 1996), 19.

[7] "Measuring Emotion," *Perception Research Services International, Inc.,* accessed February 8, 2015, http://www.prsresearch.com/prs-tools/measuring-emotion/.

[8] Martin Lindstrom, *Buyology* (New York: Crown Business, 2010), 28.

9   *Sensory Logic,* accessed February 8, 2015, http://www.sensorylogic.com/.

10 Van Praet, *Unconscious Branding,* 84.

11 Ibid., 173.

12 "What Is fMRI?" *Center for Functional MRI: UC San Diego School of Medicine,* accessed May 26, 2014, http://fmri.ucsd.edu/Research/whatisfmri.html.

    "Functional magnetic resonance imaging," *Wikipedia,* last updated 30 May 2015, accessed June 15, 2015, http://en.wikipedia.org/wiki/Functional_magnetic_resonance_imaging#Combining_with_other_methods.

13 "What Is fMRI?"

14 Margaret Talbot, "Duped," *New Yorker,* July 2, 2007, accessed April 22, 2015, http://www.newyorker.com/magazine/2007/07/02/duped.

15 Jeffrey Rosen, "The Brain on the Stand," *New York Times Magazine,* March 11, 2007, accessed February 8, 2015, http://www.nytimes.com/2007/03/11/magazine/11Neurolaw.t.html?pagewanted=6&_r=0.

    "No Lie MRI," *Wikipedia,* last updated 9 July 2014, accessed June 15, 2015, http://en.wikipedia.org/wiki/No_Lie_MRI.

16 Lindstrom, *Buyology,* 15.

17 Amber Haq, "This Is Your Brain on Advertising," *Bloomberg Business,* October 8, 2007, accessed March 7, 2015, http://www.bloomberg.com/bw/stories/2007-10-08/this-is-your-brain-on-advertisingbusinessweek-business-news-stock-market-and-financial-advice.

[18] "Benjamin Libet," *Wikipedia*, last updated May 2, 2015, accessed June 15, 2015. http://en.wikipedia.org/wiki/Benjamin_Libet.

[19] Kerri Smith, "Brain makes decisions before you even know it," *Nature*, accessed January 30, 2015, http://www.nature.com/news/2008/080411/full/news.2008.751.html.

Van Praet, *Unconscious Branding*, 5.

[20] Kathryn Hauser, "Retailers Using Science to Shape Shopping Experience," *WBZ-TV*, December 9, 2014, accessed December 10, 2014, http://boston.cbslocal.com/2014/12/09/retailers-using-science-to-shape-shopping-experience/.

[21] Hauser, "Retailers Using Science."

Reading Your Emotions and Targeting You

[1] Yuval Mor, "Emotions Analytics to Transform Human–Machine Interaction," *Huffington Post*, last updated January 23, 2014, accessed February 21, 2015, http://www.huffingtonpost.com/yuval-mor/emotions-analytics-to-tra_b_4240454.html.

[2] Rohan Joseph D'Sa, "Emotion recognition of speech signals," *RTNS*, January 15, 2003, accessed January 29, 2015, http://www.rtns.org/rohan/Emotion%20recognition%20of%20speech%20%20signals.htm.

[3] *Beyond Verbal*, accessed February 21, 2015, http://www.beyondverbal.com/.

[4] Note that some of the early documents about this company and this software used the word *deduce*. The word *infer* is probably more accurate. Deduction requires more rigid logical proofs and is often used in mathematics.

5   "Your phone says: 'Cheer up!'" *The Economist,* January 3, 2015, accessed January 5, 2015, http://www.economist.com/news/business/21637400-software-senses-how-you-are-feeling-being-pitched-gadget-makers-your-phone-says.

6   *Ginger.io,* accessed January 30, 2015, https://ginger.io/.

    Joseph Walker, "Can a Smartphone Tell if You're Depressed?" *Wall Street Journal,* January 5, 2015, accessed January 5, 2015, http://www.wsj.com/articles/can-a-smartphone-tell-if-youre-depressed-1420499238?mod=WSJ_hpp_MIDDLENextto WhatsNewsThird.

7   Reed Albergotti, "Facebook Buys Voice-Recognition Startup," *Wall Street Journal,* January 5, 2015, accessed January 5, 2015, http://www.wsj.com/articles/facebook-buys-voice-recognition-startup-1420496634?mod=WSJ_hp_LEFTWhatsNewsCollection.

8   Natasha Lomas, "Today in Creepy Privacy Policies, Samsung's Eavesdropping TV," *TechCrunch,* February 8, 2015, accessed February 8, 2015, http://techcrunch.com/2015/02/08/telescreen/.

9   Andrew Griffin, "Samsung's new smart TV policy allows company to listen in on users," *The Independent,* accessed February 9, 2015, http://www.independent.co.uk/life-style/gadgets-and-tech/news/samsungs-new-smart-tv-policy-allows-company-to-listen-in-on-users-10033012.html?icn=puff-13.

10 Ibid.

11 Keach Hagey, Shalini Ramachandran, and Daisuke Wakabayashi, "Apple Plans Web TV Service in Fall," *Wall Street Journal,* March 16, 2015, accessed March 17, 2015, http://www.wsj.

com/articles/apple-in-talks-to-launch-online-tv-service-1426555611?mod=WSJ_hp_LEFTTopStories.

12   Bruce Schneier, *Data and Goliath: The Hidden Battle to Collect Your Data and Control You* (New York: W. W. Norton, 2015), 6.

13   "Military-industrial complex," *Wikipedia,* last updated June 6, 2015, accessed June 15, 2015, http://en.wikipedia.org/wiki/Military%E2%80%93industrial_complex.

14   Jenna Wortham, "Apple Buys a Start-Up for Its Voice Technology," *New York Times,* April 29, 2010, accessed March 30, 2015, http://www.nytimes.com/2010/04/29/technology/29apple.html?_r=0.

15   *SRI International,* accessed March 30, 2015, http://www.sri.com/sites/default/files/brochures/sri-overview.pdf, 2.

16   Brad Heath, "U.S. secretly tracked billions of calls for decades," *USA Today,* April 8, 2015, accessed April 8, 2015, http://www.usatoday.com/story/news/2015/04/07/dea-bulk-telephone-surveillance-operation/70808616/.

17   Lisa A. Williams and Eliza Bliss-Moreau, "Your smartphone is looking at you—but can it read your emotions?" *The Conversation,* March 11, 2014, accessed January 17, 2015, http://theconversation.com/your-smartphone-is-looking-at-you-but-can-it-read-your-emotions-23908.

18   "Facial Action Coding System," *Wikipedia,* last updated May 5, 2015, accessed June 15, 2015, http://en.wikipedia.org/wiki/Facial_Action_Coding_System.

19   Paul Ekman, *Emotions Revealed* (New York: St. Martin's Press, 2007), 2.

20   "Silvan Tomkins," *Wikipedia,* last updated November 9,

20. 2014, accessed June 15, 2015, http://en.wikipedia.org/wiki/Silvan_Tomkins.

21. Malcolm Gladwell, "The Naked Face," *Gladwell.com,* August 5, 2002, accessed February 14, 2015, http://gladwell.com/the-naked-face/.

22. Ibid.

23. Ibid.

24. Julian Ryall, "Robots provide a personal touch at Japanese bank," *The Telegraph,* February 2, 2015, accessed February 2, 2015, http://www.telegraph.co.uk/news/worldnews/asia/japan/11384391/Robots-provide-a-personal-touch-at-Japanese-bank.html.

25. "Affectiva," *Wikipedia,* last updated January 15, 2015, accessed June 2, 2015, http://en.wikipedia.org/wiki/Affectiva.

26. *ooVoo,* accessed June 16, 2015, http://www.oovoo.com/home.aspx.

27. Elizabeth Dwoskin and Evelyn M. Rusli, "The Technology that Unmasks Your Hidden Emotions," *Wall Street Journal,* January 28, 2015, accessed January 31, 2015, http://www.wsj.com/articles/startups-see-your-face-unmask-your-emotions-1422472398?mod=LS1.

28. Michael Froimowitz Greenzeiger, Ravindra, Phulari, and Mehul K. Sanghavi, "Inferring User Mood Based on User and Group Characteristic Data," United States patent application no. 13/556023, filed July 23, 2012, *United States Patent and Trademark Office,* accessed January 17, 2015, http://appft.uspto.gov/netacgi/nph-Parser?Sect1=PTO1&Sect2=HITOFF&d=PG01&p=1&u=%2Fnetahtml%2FPTO%2Fsr

chnum.html&r=1&f=G&l=50&s1=%2220140025620%22.
PGNR.&OS=DN/20140025620&RS=DN/20140025620.

29 "Apple," *Data Center Knowledge: iNET Interactive,* accessed February 23, 2015, http://www.datacenterknowledge.com/archives/category/companies/apple.

Daniel Eran Dilger, "First Look: Apple's iCloud data center site in Reno, Nevada," *Inside Apple,* March 9, 2013, accessed February 23, 2015, http://appleinsider.com/articles/13/03/09/first-look-apples-icloud-data-center-site-in-reno-nevada.

Chris Foresman, "Apple breaks ground on mammoth Oregon data center," *Ars Technica,* October 19, 2012, accessed February 23, 2015, http://arstechnica.com/apple/2012/10/apple-breaks-ground-on-mammoth-colossal-gargantuan-oregon-data-center/.

30 Everett Rosenfeld, "Apple announces $2B global command center in Arizona," *CNBC,* February 2, 2015, accessed February 23, 2015, http://www.cnbc.com/id/102389945.

31 Yevgeniy Sverdlik, "Apple Data Center Coming to Arizona," February 2, 2015, accessed February 23, 2015, http://www.datacenterknowledge.com/archives/2015/02/02/apple-data-center-coming-to-arizona/.

32 "Utah Data Center," *Wikipedia,* last updated 13 June 2015, accessed June 15, 2015, http://en.wikipedia.org/wiki/Utah_Data_Center.

33 "MilCon Status Report—August 2014—Under Secretary of Defense for AT&L," *Office of the Under Secretary of Defense for Acquisition, Technology, and Logistics,* September 17, 2014, accessed February 22, 2015, http://www.acq.osd.mil/ie/fim/library/milcon/MILCON_EOM-AUG_Report_2014-09-17.xlsx, line 1,607.

34 Kashmir Hill, "Blueprints of NSA's Ridiculously Expensive Data Center in Utah Suggest It Holds Less Info than Thought," *Forbes*, July 24, 2013, accessed February 24, 2015, http://www.forbes.com/sites/kashmirhill/2013/07/24/blueprints-of-nsa-data-center-in-utah-suggest-its-storage-capacity-is-less-impressive-than-thought/.

35 Verlyn Klinkenborg, "Trying to Measure the Amount of Information that Humans Create," *New York Times*, November 12, 2003, accessed February 23, 2015, http://www.nytimes.com/2003/11/12/opinion/12WED4.html.

36 Marie Mawad, "Apple to Spend $1.9 Billion Building Two Europe Data Centers," *Bloomberg*, February 23, 2015, accessed February 23, 2015, http://www.bloomberg.com/news/articles/2015-02-23/apple-to-build-1-9-billion-data-centers-in-denmark-ireland.

37 Apple built a data center in Prineville, Oregon, consisting of two 338,000-square-foot buildings. *The Oregonian* speculated it would cost over $1 billion (http://arstechnica.com/apple/2012/10/apple-breaks-ground-on-mammoth-colossal-gargantuan-oregon-data-center/). In comparison, a 400,000-square-foot warehouse would cost about $7.6 million (http://www.bankrate.com/brm/news/biz/thumb/20000111.asp). Even if additional items like a cooling system, extra electrical wiring, and extra security pushed the cost up fivefold to $38 million for one 400,000-square-foot warehouses or to $76 million for two 400,000-square-foot warehouses, it would remain likely that more than 90 percent of the cost of a data center is for the electronic memory.

It may cost more to build in Europe than in the United States, since the cost of building, permits, etc., may be more in Europe. However, 90 percent or more of the cost of a data storage center is allocated for computer memory. The cost of memory has been falling for the past fifty years. In 2004, memory cost sixteen cents per megabyte. Ten years later, this dropped to less than a penny (0.8 cent) per megabyte. The cost of memory continues to drop. Therefore, Apple's newest data center is likely to hold an astronomical amount of data.

38 Geoffrey Smith, "Apple's Crazy-Expensive New Data Centers Will Be Totally Green," *Time,* February 23, 2015, accessed February 24, 2015, http://time.com/3718616/apple-data-centers-green/.

39 Mawad, "Apple to Spend $1.9 Billion."

40 Apple will likely soon have eight data centers. The locations of these include the following:

> Maiden, North Carolina — Timothy Prickett Morgan, "A peek inside Apple's iCloud data center," *The Register,* June 9, 2011, accessed March 20, 2015, http://www.theregister.co.uk/2011/06/09/apple_maiden_data_center/.
>
> Prineville, Oregon — Rich Miller, "Apple Confirms Plans for Oregon Data Center," *Data Center Knowledge,* February 21, 2012, accessed March 20, 2015, http://www.datacenterknowledge.com/archives/2012/02/21/apple-confirms-plans-for-oregon-data-center/.
>
> Newark, California — Rich Miller, "Apple Buys California Data Center," *Data Center Knowledge,* February 27, 2006, accessed March 30, 2015, http://www.datacenterknowledge.com/archives/2006/02/27/apple-buys-california-data-center/.

Santa Clara, California — Rich Miller, "Apple Adding Data Center in Silicon Valley," *Data Center Knowledge,* May 18, 2011, accessed March 30, 2015, http://www.datacenterknowledge.com/archives/2011/05/18/apple-adding-data-center-in-silicon-valley/.

Reno, Nevada — Yevgeniy Sverdlik, "UPDATED: Apple's Reno Data Center Expansion Marches On," *Data Center Knowledge,* August 8, 2014, accessed March 30, 2015, http://www.datacenterknowledge.com/archives/2014/08/08/apple-data-center-expansion-in-reno-marches-on/.

Mesa, Arizona — Daisuke Wakabayashi, "Apple to Build Data Command Center in Arizona," *Wall Street Journal,* February 2, 2015, accessed March 20, 2015, http://blogs.wsj.com/digits/2015/02/02/apple-to-build-data-command-center-in-arizona/.

Apple plans to build two data centers in Europe. See Lisa Fleisher, "Apple to Spend Nearly $2 Billion on New European Data Centers," *Wall Street Journal,* February 23, 2015, accessed March 30, 2015, http://www.wsj.com/articles/apple-to-invest-1-9-billion-in-european-data-centers-1424685191.

Note that the cost of data storage is falling rapidly. Therefore, by the time the two data storage centers are finished in Europe, each may store about the same as the Utah Data Center, even though the cost of each may be less than the Utah Data Center.

41 Google Data Centers, accessed April 21, 2015, http://www.google.com/about/datacenters/inside/locations/index.html.

42 Google Data Centers, accessed April 21, 2015, http://www.google.com/about/datacenters/inside/locations/mayes-county/.

43 Google Data Centers, accessed April 21, 2015, http://www.google.com/about/datacenters/inside/locations/berkeley-county/.

44 Google Data Centers, accessed April 21, 2015, http://www.google.com/about/datacenters/inside/locations/council-bluffs/.

45 "Facebook Data Center FAQ," *Data Center Knowledge*, accessed April 21, 2015, http://www.datacenterknowledge.com/the-facebook-data-center-faq/.

46 Mark Hosenball, "NSA chief says Snowden leaked up to 200,000 secret documents," *Reuters*, November 14, 2013, accessed April 3, 2015, http://www.reuters.com/article/2013/11/14/us-usa-security-nsa-idUSBRE9AD19B20131114.

47 Alina Selyukh and Greg Savoy, "Protesters march in Washington against NSA spying," ed. Peter Cooney, *Reuters*, October 26, 2013, accessed April 3, 2015, http://www.reuters.com/article/2013/10/27/us-usa-security-protest-idUSBRE99P0B420131027.

48 Hosenball, "NSA chief says Snowden leaked up to 200,000 secret documents."

49 *Der Spiegel*, accessed April 3, 2015, http://www.spiegel.de/international/.

Playing with Emotions Is Very Profitable

1. Adam D. I. Kramer, Jamie E. Guillory, and Jeffrey T. Hancock, "Experimental evidence of massive-scale emotional contagion through social networks, *Proceedings of the National Academy of Sciences of the United States of America,* March 25, 2014, accessed February 13, 2015, http://www.pnas.org/content/111/24/8788.full.pdf.

2. Vindu Goel, "Facebook Tinkers with Users' Emotions in News Feed Experiment, Stirring Outcry," *New York Times,* June 29, 2014, accessed February 13, 2015, http://www.nytimes.com/2014/06/30/technology/facebook-tinkers-with-users-emotions-in-news-feed-experiment-stirring-outcry.html.

3. Lori Andrews, *I Know Who You Are and I Saw What You Did* (New York: Simon and Schuster, Inc.), Kindle edition. (See p. 5, location 209 of 7,414.)

4. Michael Fertik, "Federal Trade Commission Ponders Digital Privacy," *MichaelFertik.com,* December 12, 2009, accessed February 21, 2015, http://michaelfertik.com/news/federal-trade-commission-ponders-digital-privacy/.

5. Robert Heath, *Seducing the Subconscious: The Psychology of Emotional Influence in Advertising* (Malden, MA: Wiley Publishing, 2012).

6. Jean Kilbourne, *Deadly Persuasions* (New York: The Free Press, 1999), 27.

7. "Marlboro (cigarette)," *Wikipedia,* last updated June 12, 2015, accessed June 15, 2015, http://en.wikipedia.org/wiki/Marlboro_%28cigarette%29.

8   James B. Twitchell, *Twenty Ads that Shook the World* (New York: Crown Publishers, 2000), 129.

9   Ellen Shapiro, *The 100 Most Influential People Who Never Lived,* ed. Kelly Knauer (New York: Time Books, 2013).

10 "Marlboro Man," *Wikipedia,* last updated May 18, 2015, accessed June 16, 2015, https://en.wikipedia.org/wiki/Marlboro_Man.

11 Twitchell, *Twenty Ads that Shook the World.*

12 Gerald Zaltman, *How the Consumers Think: Essential Insight into the Mind of the Market* (Boston: Harvard Business School Publishing, 2003), 71.

Manda Mahoney, "The Subconscious Mind of the Consumer (And How to Reach It)," *Harvard Business School Working Knowledge,* accessed April 24, 2013, http://hbswk.hbs.edu/item/3246.html.

13 Roger Dooley, "Neuromarketing: Where Brain Science and Marketing Meet," *Neuroscience Marketing Blog,* accessed April 24, 2014, http://www.neurosciencemarketing.com/blog.

14 Roger Dooley, *Brainfluence* (Hoboken: John Wiley and Son, 2012), 2.

15 "About Jean: Full Bio," *Jean Kilbourne,* accessed February 13, 2015, http://www.jeankilbourne.com/bio/.

16 Kilbourne, *Deadly Persuasions.*

17 Patrick Renvoise and Christopher Morin, "Foreword," in Robert Bishop, *Neuromarketing* (Dallas: Thomas Nelson, 2007), viii.

Manipulating with Fear

1. Edna B. Foa and Michael J. Kozak, *Emotional Processing of Fear: Exposure to Corrective Information, Psychological Bulletin* 99, no. 1 (1986): 20–35, accessed January 2, 2015, http://www.personal.kent.edu/~dfresco/CBT_Readings/foa_&_kozak_1986.pdf.

2. Coby Ben-Simhon, "Queen of broken hearts," *Haaretz*, August 6, 2010, accessed February 12, 2015, http://www.haaretz.com/weekend/magazine/queen-of-broken-hearts-1.306416.

3. "Fear," *Wikipedia*, last updated 9 June 2015, accessed June 15, 2015, http://en.wikipedia.org/wiki/Fear.

4. Edna B. Foa, Elizabeth A. Hembree, and Barbara Olawov Rothbaum, *Treatments that Work Series: Prolonged Exposure Therapy for PTSD, Emotional Processing of Traumatic Events, Therapist Guide* (New York: Oxford University Press, 2007), 13.

5. Daniel Gardner, *Risk: The Science and Politics of Fear* (London: Virgin Books, 2009), 126–8.

6. Marc Andrews, Matthijs van Leeuwen, and Rick van Baaren, *Hidden Persuasion: 33 Psychological Influence Techniques in Advertising* (Amsterdam: BIS Publishers, 2013), 158.

7. Gardner, *Risk*, 149–260.

   Daniel Gardner, *Science of Fear: How the Culture of Fear Manipulates People's Brains* (New York: Plume Books, 2009), 65–66.

8. Gardner, *Science of Fear*, 10.

9. Ibid., 13.

The "Belongings Test": What You Buy and Where You Go Reveals More than You Realize

1. Vance Packard, *The Hidden Persuaders* (New York: David McKay Company, 1957), 46–56.

2. Sam Gosling, *Snoop: What Your Stuff Says about You* (New York: Basic Books, 2008), 75.

3. *Mandala Consulting,* accessed February 12, 2015, http://www.mandalaconsulting.co.za/Documents/Reviews/Review%201.pdf.

4. Ellen Gamerman, "When the Art Is Watching You," *Wall Street Journal,* December 11, 2014, accessed February 25, 2015, http://www.wsj.com/articles/when-the-art-is-watching-you-1418338759?KEYWORDS=data+mining+medical+information.

5. Nick Saint, "Google CEO: 'We Know Where You Are. We Know Where You've Been. We Can More or Less Know What You're Thinking About,'" *Business Insider,* October 4, 2010, accessed March 25, 2015, http://www.businessinsider.com/eric-schmidt-we-know-where-you-are-we-know-where-youve-been-we-can-more-or-less-know-what-youre-thinking-about-2010-10.

6. Charles Duhigg, "How Companies Learn Your Secrets," *New York Times,* February 19, 2012, accessed April 23, 2015, http://www.nytimes.com/2012/02/19/magazine/shopping-habits.html?pagewanted=all&_r=0.

7. Kashmir Hill, "How Target Figured Out a Teen Girl Was Pregnant before Her Father Did," *Forbes,* February 16, 2012, accessed February 12, 2015, http://www.forbes.com/sites/kashmirhill/2012/02/16/how-target-figured-out-a-teen-girl-was-pregnant-before-her-father-did/.

Duhigg, "How Companies Learn Your Secrets."

Bruce Schneier, *Data and Goliath: The Hidden Battle to Collect Your Data and Control You* (New York: W. W. Norton, 2015), 33.

8 "Of sensors and sensibility," *The Economist*, April 2, 2015, accessed April 8, 2015, http://www.economist.com/news/business-and-finance/21647715-connected-devices-home-are-becoming-more-widespread-sensors-and-sensibility.

9 Andrea Chang, "Facebook to roll out payments feature on Messenger," *Los Angeles Times*, March 17, 2015, accessed March 17, 2015, http://www.latimes.com/business/technology/la-fi-tn-facebook-messenger-send-money-20150317-story.html?track=rss.

"Unfriending Cash," *The Economist*, March 21, 2015, accessed April 8, 2015, http://www.economist.com/news/finance-and-economics/21646802-facebook-enters-booming-market-mobile-payments-unfriending-cash.

10 United States patent application no. 20130144789, "Method and System for Managing Credits via a Mobile Device," filed January 29, 2013, *United States Patent and Trademark Office*, accessed March 31, 2015, http://appft.uspto.gov/netacgi/nph-Parser?Sect1=PTO2&Sect2=HITOFF&p=1&u=%2Fnetahtml%2FPTO%2Fsearch-adv.html&r=1&f=G&l=50&d=PG01&S1=%28%22Managing+Credits%22.TTL.%29&OS=TTL/%22+Managing+Credits%22&RS=TTL/%22+Managing+Credits%22.

11 David Alton Clark, "Apple: About to Show You the iMoney?" *Seeking Alpha*, June 7, 2013, accessed February 12, 2015, http://seekingalpha.com/article/1488192-apple-about-to-show-you-the-imoney.

12. Robin Sidel and Daisuke Wakabayashi, "Apple Pay Stung by Low-Tech Fraudsters," *Wall Street Journal,* March 5, 2015, accessed March 5, 2015, http://www.wsj.com/articles/apple-pay-stung-bylow-techfraudsters-1425603036?mod=WSJ_hp_LEFTTopStories.

13. Devlin Barrett, "CIA Aided Program to Spy on U.S. Cellphones," *Wall Street Journal,* March 10, 2015, accessed March 12, 2015, http://www.wsj.com/articles/cia-gave-justice-department-secret-phone-scanning-technology-1426009924?mod=trending_now_10.

14. Emily Steel, "Companies scramble for consumer data," *Financial Times,* June 12, 2013, accessed March 3, 2015, http://www.ft.com/intl/cms/s/0/f0b6edc0-d342-11e2-b3ff-00144feab7de.html#axzz2W71Wg60o (subscription required).

15. Anton Troianovski, "Phone Firms Sell Data on Customers," *Wall Street Journal,* May 21, 2013, accessed March 12, 2015, http://www.wsj.com/articles/SB10001424127887323463704578497153556847658.

16. "In the Matter of Snapchat, Inc.," *Federal Trade Commission,* accessed March 10, 2015, https://www.ftc.gov/system/files/documents/cases/140508snapchatcmpt.pdf, count 2, pp. 5 and 6.

17. "Text of the Location Privacy Protection Act of 2014," *GovTrack,* accessed March 9, 2015, https://www.govtrack.us/congress/bills/113/s2171/text.

18. "Smartphone Apps Covertly Report Your Location Data" Uploaded on April 3, 2015, Cyber Security Intelligence, accessed June 12, 2015 http://www.cybersecurityintelligence.com/blog/smartphone-apps-covertly-report-your-location-data--189.html.

19 Anton Troianovski, "Phone Firms Sell Data on Customers," *Wall Steet Journal*, May 21, 2013 accessed June 12, 2015, http://www.wsj.com/articles/SB10001424127887323463704578497153556847658.

20 Sara Silverstein, "These Animated Charts Tell You Everything about Uber Prices in 21 Cities," *Business Insider*, October 16, 2014, accessed March 3, 2015, http://www.businessinsider.com/uber-vs-taxi-pricing-by-city-2014-10.

21 Ren Lu, "Making a Bayesian Model to Infer Uber Rider Destinations," *Uber Blog*, September 2, 2014, accessed March 2, 2015, http://blog.uber.com/passenger-destinations.

Bradley Voytek, "Mapping the San Franciscome," *Uber Blog*, January 9, 2012, accessed February 13, 2015, http://blog.uber.com/2012/01/09/uberdata-san-franciscomics/ or http://blog.uber.com/2012/01/09/uberdata-san-franciscomics/.

22 Schneier, *Data and Goliath*, 3.

23 Tomio Geron, "Uber Confirms $258 Million from Google Ventures, TPG, Looks to On-Demand Future," *Forbes*, August 23, 2013, accessed March 3, 2015, http://www.forbes.com/sites/tomiogeron/2013/08/23/uber-confirms-258-million-from-google-ventures-tpg-looks-to-on-demand-future/.

24 Scott Austin, Chris Canipe, and Sarah Slobin, "The Billion Dollar Startup Club," *Wall Street Journal*, accessed February 19, 2015, http://graphics.wsj.com/billion-dollar-club/?co=Uber.

25 Tim Higgins, "Apple Wants to Start Producing Cars as Soon as 2020," *Bloomberg*, February 19, 2015, accessed February 20, 2015, http://www.bloomberg.com/news/articles/2015-02-19/apple-said-to-be-targeting-car-production-as-soon-as-2020?ho

otPostID=9f3917f4905f086a568d81c5d794101a.

26 John Markoff, "Google's Next Phase in Driverless Cars: No Steering Wheel or Brake Pedals," *New York Times,* May 27, 2014, accessed February 20, 2015, http://www.nytimes.com/2014/05/28/technology/googles-next-phase-in-driverless-cars-no-brakes-or-steering-wheel.html?_r=0.

27 Claudia Assis, "The list of electric-car projects to rival Tesla is getting longer," *Market Watch,* February 20, 2015, accessed February 20, 2015, http://www.marketwatch.com/story/the-list-of-electric-car-projects-to-rival-tesla-is-getting-longer-2015-02-20?siteid=rss&rss=1.

28 Mia De Graaf, "Head of Google's secretive X lab defends Glass headset against privacy campaigners—and explains why engineers now wear fluffy socks to work," *Daily Mail,* last updated: 9:30 p.m. EST, March 18, 2015, accessed March 18, 2015, http://www.dailymail.co.uk/sciencetech/article-2999644/Head-Google-s-secretive-X-lab-defends-firm-s-Glass-headset-against-privacy-campaigners-explains-engineers-wear-fluffy-socks-work.html.

29 Emily Steel, "Companies scramble for consumer data."

Big Data Gets to Know You, One Very Small Piece of Information at a Time

[1] Piecing together a person's psychological profile is not like reading a book. When you read this book, you see that it is a sequential, or linear, process of gaining understanding. Your mind combines individual letters to make words. Then you read one word after another. Each additional word builds on the understanding of the previously read words.

If I have done my job well as the author of this book, then the words will be arranged neatly into sentences and paragraphs to provide you with a framework so your mind can put the words into a meaningful order. This makes understanding much easier than if I randomly threw thousands of individual letters at you.

Such is not the case with DNA sequencing. When a person's DNA is sequenced, it is not analyzed by reading all of that person's genetic material (genome) from one end to the other in a sequential manner (which is how you are reading the letters and words in this book). DNA strands are too long to make that possible. Perhaps in the future, some machine will be able to read DNA sequentially from one end to the other. For now, however, DNA has to be cut into short segments, analyzed, and then electronically reassembled in order to be read.

This book contains about two hundred and eighty thousand characters, including spaces, punctuation marks, the glossary, the bibliography, and the notes. Imagine that you randomly cut up thousands of exact copies of this book into millions of different segments, each ranging from a hundred to three hundred letters (spaces and punctuation marks included) in length. Assume that none of the segments is the same. Also assume that many of these segments have portions of the segment that could be cross-matched with parts of the DNA of other segments.

It is important to note that the product described by using this analogy is not like a jigsaw puzzle, in which the pieces are matched by their edges. The pieces are matched by overlapping information. The overlapping aspect of small pieces of information allows a meaningful connection to other small pieces of information. In this manner, millions of small, seemingly disconnected pieces of information can be neatly connected.

To make sense of these millions of randomly cut segments, you have to reconstruct the original manuscript of this book by cross-matching the various segments together. In order to match the segments, you look for sequences of letters at the beginning and at the end of these thousand-letter segments that match similar sequences in other randomly cut segments. Some of the segments contain short sequences of letters (or words) that match partial portions of other segments. In the long and laborious process of matching portions of the one hundred to three hundred letter segments with millions of other segments, eventually you could construct a manuscript and have a high degree of confidence that it matched the original manuscript. This is a mind-numbing task that few people would attempt. However, computers can perform tasks similar to this one very rapidly.

Nucleotides, or base pairs, can be compared to individual letters in the words of a manuscript. In a manner similar to cutting this manuscript into segments, long pieces of DNA can be randomly cut up with enzymes (Stuart M. Brown, "Introduction to DNA Sequencing," in *Next-Generation DNA Sequencing Infomatics,* ed. Stuart M. Brown [Cold Springs Harbor, NY: Cold Springs Harbor Laboratory Press, 2013], 5) into millions of separate segments, each containing around a hundred nucleotides, or base pairs.

Computers with sophisticated software look for areas where the sequences of base pairs overlap and then reconstruct entire DNA sequences by cross-matching the individual segments (Efstratios Eesstrathiadis and Eric R. Peskin, "High Performance Computing in DNA Sequencing Informations," in *Next-Generation DNA Sequencing Infomatics,* 195).

The important point to remember is that a very small segment of DNA by itself means almost nothing. However, when cross-matched against millions other pieces, the entire original set of genetic material can be read in amazing detail, often with over 99 percent accuracy (Stuart M. Brown, "Introduction to DNA Sequencing," in *Next-Generation DNA Sequencing Infomatics,* 21).

Here is an important difference. In one example, the whole is cut into pieces to examine its small segments. Then the many individual segments are electronically reassembled. In the other case, millions of pieces of information are gathered and electronically assembled to create a person's psychological dossier. Stated differently, a person's DNA can be analyzed by cutting it into millions of pieces to be analyzed. However, a person's psychological profile can be assembled by gathering millions of pieces of information about that person that are found on the Internet and on his or her smartphone. Just like the individual small segments of DNA are meaningless by themselves, small and seemingly isolated pieces of information about a person may seem meaningless by themselves. However, millions of small pieces of DNA can be analyzed and reassembled electronically to reveal a person's entire genetic code.

In a similar manner, a person's psychological profile can be established from thousands or millions of very small pieces of information that by themselves are almost insignificant and seemingly incongruous. This psychological profile of a person's emotions, moods, aspirations, and fears reconstructed from a person's online behavior is more likely to be accurate than if the data gatherers read all of the person's diaries and subjected the person to weeks of interviews with psychologists, psychiatrists, and anthropologists. Much like how computers reconstruct a person's DNA from millions of small segments, computers can reconstruct a person's psychological profile in a mathematic process.

If you want to know more, I refer you to *Introduction to Mathematical Sociology* by Phillip Bonacich and Philp Lu (Princeton: Princeton University Press, 2012), and to *Introduction to Mathematical Sociology* by James S. Coleman (London: Collier-Macmillan, 1964).

2   Mary Wollstonecraft Shelley, *Frankenstein; or, The Modern Prometheus* (Garden City, NY: Halcyon House, 2010), 202-3.

> Soon after my arrival in the hovel I discovered some papers in the pocket of the dress which I had taken from your laboratory. At first I had neglected them; but now that I was able to decipher the characters in which they were written, I began to study them with diligence. It was your journal of the four months that preceded my creation. You minutely described in these papers every step you took in the progress of your work; this history was mingled with accounts of domestic occurrences. You doubtless recollect these papers. Here they are. Everything is related in them which bears reference to my accursed origin; the whole detail of that series of disgusting circumstances which produced it is set in view; the minutest description of my odious and loathsome person is given, in language which painted your own horrors and rendered mine indelible. I sickened as I read.

3   Amendment 4 to Form S-1, "Registration Statement," *Securities and Exchange Commission,* July 26, 2004, accessed February 21, 2015, http://www.sec.gov/Archives/edgar/data/1288776/000119312504124025/ds1a.htm, 12.

4   Google, Inc., Gmail litigation, case no. 5:13-md-02430-LHK, District Court of Northern California, reported by *Consumer Watchdog,* filed on September 5, 2013, accessed February 21, 2015, http://www.consumerwatchdog.org/resources/googlemotion061313.pdf.

5   Jonathan Stempel, "Google won't face email privacy class action," *Reuters,* March 19, 2014, accessed February 21, 2015, http://www.reuters.com/article/2014/03/19/us-google-gmail-lawsuit-idUSBREA2I13G20140319.

6   Google, Inc., Gmail litigation, case no. 13-MD-02430-LHK, United States District Court, Northern District of California, San Jose Division, March 18, 2014.

"users 'give Google (and those [whom Google] work[s] with) a worldwide license to use ... [and] create derivative works (such as those resulting from translations, adaptations or other changes we make so that your content works better with our Services) ... and distribute such content'" (p. 6).

Google updated the policy to state that the collected information would be used to "provide, maintain, protect, and improve our services (including advertising services) and develop new services" (p. 7).

7   Google, Inc., Gmail litigation, case no. 13-MD-02430-LHK.

8   "Google to stop scanning youth accounts in education suite," *Al Jazeera* and the Associated Press, April 30, 2014, accessed April

30, 2014, http://america.aljazeera.com/articles/2014/4/30/google-scanning-emailkids.html.

9   Alexei Oreskovic, "Google spells out email scanning practices in new terms of service," *Reuters,* April 14, 2014, accessed March 5, 2015, http://www.reuters.com/article/2014/04/14/google-email-idUSL2N0N61MT20140414.

10 Although Google had been reading Gmail since 2004, it first indicated to users that it was doing so in 2007—and then only very obliquely—in "section of the 2007 TOS [terms of service] provided that "some of the

Services are supported by advertising revenue and may display advertisements and promotions" and that "these advertisements may be content-based to the content information stored on the

Services, queries made through the Service or other information." Google, Inc., Gmail litigation, case no. 13-MD-02430-LHK, United States District Court, Northern District of California, San Jose Division, March 18, 2014, signed by Lucy H. Koh, United States district judge, accessed December 15, 2014.

"Whether Plaintiffs Consented to Automated Scanning Pursuant to § 2511(2)(d) Is a Controlling Question of Law" (pages 19–20) and "Substantial Ground for Difference of Opinion Exists as to Whether Plaintiffs

Consented to Automated Scanning Pursuant to § 2511(2)(d)" (pages 20–24). Google, Inc., Gmail litigation, case no. 13-MD-02430-LHK, United States District Court, Northern District of California, San Jose Division, October 9, 2013, signed by Kathleen M. Sullivan, attorney for the defendant Google, Inc., accessed December 15, 2014.

11. "What Is Watson?" *IBM,* accessed March 5, 2015, http://www.ibm.com/smarterplanet/us/en/ibmwatson/what-is-watson.html.

12. Joab Jackson, "IBM Watson Vanquishes Human Jeopardy Foes," *PCWorld,* February 16, 2011, accessed December 15, 2014, http://www.pcworld.com/article/219893/ibm_watson_vanquishes_human_jeopardy_foes.html.

13. "A Cure for the Big Blues," *The Economist,* January 11, 2014, accessed January 14, 2014, http://www.economist.com/news/business/21593489-technology-giant-asks-watson-get-it-growing-again-cure-big-blues.

14. Chris Crum, "Google Outlines What It's Doing to Protect Your Data from the Government," *WebProNews,* January 28, 2013, accessed February 21, 2015, http://www.webpronews.com/google-outlines-what-its-doing-to-protect-your-data-from-the-government-2013-01.

    "Google Will Fight Government Over Access to Your Emails," *Reuters,* January 28, 2013, accessed February 21, 2015, http://www.reuters.com/article/2013/01/28/us-usa-privacy-google-idUSBRE90R17120130128.

    "How does Google protect my privacy and keep my information secure?" *Google,* accessed February 21, 2015, http://www.google.com/policies/faq/.

    "Gonzales v. Google Inc., Filing 12," *Justia,* accessed February 21, 2015, http://docs.justia.com/cases/federal/district-courts/california/candce/5:2006mc80006/175448/12/.

15. "Transparency and secrecy: Never mind the warrants," *The Economist,* January 25, 2013, accessed March 20, 2015, http://

www.economist.com/blogs/democracyinamerica/2013/01/transparency-and-secrecy-0.

"How does Google protect my privacy and keep my information secure?"

16   Amendment 4 to Form S-1, 8.

17   Charlie Osborne, "Google launches quantum processor, artificial intelligence project," *ZDNet,* September 3, 2014, accessed March 19, 2015, http://www.zdnet.com/article/google-launches-quantum-processor-artificial-intelligence-project/.

"AI research at Google," *Artificial Brains,* accessed March 19, 2015, http://www.artificialbrains.com/google.

18   Joel Rosenblatt, "Google Wants E-Mail Scanning Information Blocked," *Bloomberg,* March 14, 2014, accessed February 21, 2015, http://www.bloomberg.com/news/2014-03-14/google-wants-e-mail-scanning-information-blocked.html.

19   Ibid.

20   Ben Rossi, "Facebook DOES collect the text you decided against posting," *Information Age,* April 8, 2015, accessed April 22, 2015, http://www.information-age.com/technology/information-management/123459286/facebook-does-collect-text-you-decided-against-posting.

21   "*Minority Report* (film)," *Wikipedia,* last updated June 11, 2015, accessed June 15, 2015. http://en.wikipedia.org/wiki/Minority_Report_%28film%29#cite_ref-ebertroeper_159-1.

22   *Predictive Analytics World,* accessed March 18, 2015, http://www.predictiveanalyticsworld.com/.

PredictiveAnalytics, accessed March 18, 2015, http://predictiveanalytics.org/.

23. "Collecting private information: Uses and abuses: A computer-security expert weighs up the costs and benefits of collecting masses of personal data," *The Economist,* April 4, 2015, accessed April 19, 2015, http://www.economist.com/news/books-and-arts/21647595-computer-security-expert-weighs-up-costs-and-benefits-collecting-masses.

24. Dana Mattioli, "On Orbitz, Mac Users Steered to Pricier Hotels," *Wall Street Journal,* August 23, 2012, accessed April 19, 2015, http://www.wsj.com/articles/SB10001424052702304458604577488822667325882.

"Anonymous," but Very Well-Known

1. "Massachusetts provider settles HIPAA case for $1.5 million," *Health and Human Services,* accessed April 14, 2015, http://www.hhs.gov/ocr/privacy/hipaa/enforcement/examples/meei-agreement.html.

   Marcy Wilder, "Alaska Medicaid Settles HIPAA Security Rule Violations for $1.7 Million," *Hogan Lovells,* accessed April 14, 2015, http://www.hldataprotection.com/2012/06/articles/health-privacy-hipaa/alaska-medicaid-settles-hipaa-security-rule-violations-for-17-million/.

   Greg Masters, "Rite Aid to pay $1 million fine for HIPAA violation," *SC Magazine,* accessed April 14, 2015, http://www.scmagazine.com/rite-aid-to-pay-1-million-fine-for-hipaa-violation/article/175729/.

2. Jordan Robertson, "Your Medical Records Are for Sale,"

Bloomberg, August 8, 2013, accessed April 7, 2015, http://www.bloomberg.com/bw/articles/2013-08-08/your-medical-records-are-for-sale.

3   Robin Respaut and Lucas Iberico Lozada, "'Slicing and dicing': How some U.S. firms could win big in 2016 elections," *Reuters,* April 14, 2015, accessed April 14, 2015, http://www.reuters.com/article/2015/04/14/us-usa-election-data-idUSKBN0N509O20150414.

4   "Protecting Consumer Privacy in an Era of Rapid Change," *Federal Trade Commission,* December 2010, accessed April 2, 2013, http://www.ftc.gov/os/2010/12/101201privacyreport.pdf, 35.

5   "What information does Facebook get when I visit a site with the Like button or another social plugin?" *Facebook,* accessed June 14, 2013, http://www.facebook.com/help/186325668085084/.

6   "Path will protect private user data with 'hashing' in next release," *The Verge,* accessed October 6, 2012, http://www.theverge.com/2012/3/8/2855189/path-will-protect-private-user-data-with-hashing-in-next-release.

7   Ed Felton, "Does Hashing Make Data 'Anonymous'?" *Tech@FTC,* accessed July 14, 2013, http://techatftc.wordpress.com/2012/04/22/does-hashing-make-data-anonymous/.

8   To fit the scenario described in this book, this illustration has been modified from one provided at "Cryptographic hash function," *Wikipedia,* last updated June 3, 2015, accessed February 25, 2015, http://en.wikipedia.org/wiki/Cryptographic_hash_function.

9   "MD5 Generator," accessed February 26, 2015, http://md5generator.net/?q.

Find an online hasher, such as the MD5 Generator (MD5 is an acronym for "Message-Digest 5"). Type in the name John Doe. The algorithm converts it to 4c2a904bafba06591225113ad17b5cec. Alternatively, enter the numeral 100. The algorithm returns a hash string of f899139df5e1059396431415e770c6dd. In this example, the MD5 algorithm generates a 32-character string in hexadecimal characters. Hexadecimal may sound complicated, but it is not. Instead of using only numerals, like 1 through 9, or only letters, this system uses 16 characters, 10 numbers, and 6 letters (namely, 1, 2, 3, 4, 5, 6, 7, 8, 9, 0, A, B, C, D, E, and F) ("Hexadecimal," *Wikipedia,* accessed February 25, 2015, http://en.wikipedia.org/wiki/Hexadecimal). Hexadecimals are more effective for computers to encode than are numbers or letters. Once converted, an entry can only be converted back to the original letters or numbers by using the same algorithm. In this case, it is an MD5 cracker tool. Once information such as your name or number is converted into the MD5 hash code, you can send it to someone else. Then that person can use the same algorithm to convert it back to the original form and get "John Doe" and "100."

[10] "Facebook and Datalogix," *Internet Privacy Information Center (EPIC),* accessed February 25, 2015, http://epic.org/privacy/facebook/facebook_and_datalogix.html.

"Facebook's Retail Data Mining Already Being Questioned by Privacy Groups," *Informed Planet,* accessed February 25, 2015, http://informedplanet.org/blog/26/facebook%E2%80%99s-retail-data-mining-already-being-questioned-by-privacy-groups/comment/.

[11] Felten, "Does Hashing Make Data 'Anonymous'?"

Timothy B. Lee, "Privacy groups seek investigation of Facebook's retail data sharing," *Ars Technica,* September 27, 2012, accessed February 25, 2015, http://arstechnica.com/tech-policy/2012/09/privacy-groups-seek-investigation-of-facebooks-retail-data-sharing/.

12   Simon Singh, *The Code Book* (New York: Anchor Books, 1999).

13   "Data Fusion: Information of the World Unite! Privacy in an Age of Terabytes and Terror," *Scientific American* 299, no. 3 (September 2008): 86.

14   "Record linkage" redirected from "Identity resolution." *Wikipedia.* Last updated June 4, 2015 Accessed June 15, 2015. https://en.wikipedia.org/wiki/Record_linkage#Identity_resolution.

15   "Hash function: Finding similar records," *Wikipedia,* last updated June 11, 2015, accessed June 15, 2015, http://en.wikipedia.org/wiki/Hash_function#Finding_similar_records.

16   "Jeff Jonas," *IBM,* accessed March 21, 2015, http://www-03.ibm.com/press/us/en/biography/40087.wss.

17   "Data Fusion."

18   "Google Policies & Principles: Anonymous identifier," *Google,* accessed June 15, 2013, http://www.google.com/policies/privacy/key-terms/#toc-terms-server-logs.

19   "IP address,"." *Wikipedia.* last updated 23 May 23, 2015, Accessed June 15, 2015, https://en.wikipedia.org/wiki/IP_address.

20   Nate Anderson, "Why Google keeps your data forever, tracks you with ads," *Ars Technica,* March 8, 2010, accessed March 5, 2015, http://arstechnica.com/tech-policy/2010/03/google-keeps-your-data-to-learn-from-good-guys-fight-off-bad-guys/.

"Understanding TCP/IP addressing and subnetting basics," *Microsoft,* accessed March 5, 2015, http://support.microsoft.com/kb/164015.

21   Anderson, "Why Google keeps your data forever."

22   "Personally identifiable information," *Wikipedia,* last updated June 8, 2015, accessed June 15, 2015. http://en.wikipedia.org/wiki/Personally_identifiable_information.

23   "Protecting Consumer Privacy in an Era of Rapid Change," 35.

24   This is meant to represent a hypothetical person. Any resemblance to a real person is coincidental and not intentional.

25   Kathy Smith and Kathleen Smith are fictitious people. Any resemblance to a real person is coincidental and not intentional.

26   Thomas Oscherwitz, "Why millions of U.S. citizens have multiple Social Security numbers," *The Last Watchdog,* September 14, 2010, accessed March 6, 2015, http://lastwatchdog.com/millions-u-s-citizens-multiple-social-security-numbers/.

Matt Cover, "Government Gave 4,317 Aliens 2 Social Security Numbers a Piece," *CNS News,* January 29, 2013, accessed March 6, 2015, http://cnsnews.com/news/article/government-gave-4317-aliens-2-social-security-numbers-piece.

Carl P. Brown, "Social Security Numbers—Use & Abuse" (originally published in *Bankers' Hotline* 1, no. 1 [January 1990]), *Bankers Online,* accessed January 7, 2015, http://www.bankersonline.com/articles/bhv01n01/bhv01n01a6.html.

Susan Jones, "IG Audit: 6.5 Million People with Active Social Security Numbers Are 112 or Older," *CNS News,* March 9, 2015, accessed March 10, 2015, http://www.cnsnews.com/news/

article/susan-jones/ig-audit-65-million-people-active-social-security-numbers-are-112-or-older.

27 Bill Meyer, "How many Americans' Social Security numbers were officially duplicated for Pacific islanders?" *Cleveland.com (Plain Dealer Publishing Co.),* August 18, 2009, accessed March 6, 2015, http://www.cleveland.com/nation/index.ssf/2009/08/how_many_americans_social_secu.html.

28 "The Value of Our Digital Identity," *The Boston Consulting Group,* accessed March 6, 2015, http://www.lgi.com/PDF/public-policy/The-Value-of-Our-Digital-Identity.pdf.

"Digital identity," *Wikipedia,* last updated June 12, 2015, accessed June 15, 2015, http://en.wikipedia.org/wiki/Digital_identity.

29 Scott Thurm and Yukari Iwatani Kane, "Your Apps Are Watching You," *Wall Street Journal,* accessed March 6, 2015, http://online.wsj.com/article/SB10001424052748704694004576020083703574602.html?mod=WSJ_WhatTheyKnow2010_WhatsNews_4_2_Right_Summaries.

30 Julia Angwin, "The Web's New Gold Mine: Your Secrets," *Wall Street Journal,* July 30, 2010, accessed March 6, 2015, http://online.wsj.com/article/SB10001424052748703940904575395073512989404.html?mod=WSJ_WhatTheyKnow2010_WhatsNews_4_2_Left.

31 "Digital identity," *Wikipedia.*

32 "Rethinking Personal Data," *World News Inc.,* accessed July 13, 2013, http://wn.com/wef_rethinking_personal_data.

33 Julia Angwin and Jeremy Singer-Vine, "Selling You on Facebook," *Wall Street Journal,* April 7, 2012, accessed March 8, 2015,

http://online.wsj.com/article/SB10001424052702303302504577327744009046230.html?mod=WSJ_WhatTheyKnowPrivacy_MIDDLETopMiniLeadStory.

34 Bruce Schneier, *Data and Goliath: The Hidden Battle to Collect Your Data and Control You* (New York: W. W. Norton, 2015), 33–45.

The Genius of Facebook

1 Nicholas Carlson, "At Last—The Full Story of How Facebook Was Founded," *Business Insider,* March 5, 2010, accessed January 8, 2015, http://www.businessinsider.com/how-facebook-was-founded-2010-3?op=1.

Ben Mezrich, *The Accidental Billionaires: The Founding of Facebook: A Tale of Sex, Money, Genius, and Betrayal* (New York: Doubleday, 2009).

2 Katharine A. Kaplan, "Facemash Creator Survives Ad Board," *The Harvard Crimson,* November 19, 2003, accessed January 8, 2015, http://www.thecrimson.com/article/2003/11/19/facemash-creator-survives-ad-board-the/.

3 Nicholas Carlson, "In 2004, Mark Zuckerberg Broke into a Facebook User's Private Email Account," *Business Insider,* March 5, 2010, accessed January 8, 2015, http://www.businessinsider.com/how-mark-zuckerberg-hacked-into-the-harvard-crimson-2010-3.

4 Ibid.

5 Dominic Rushe, "Facebook IPO sees Winklevoss twins heading for $300m fortune," *The Guardian,* February 2, 2012, accessed

January 8, 2015, http://www.theguardian.com/technology/2012/feb/02/facebook-ipo-winklevoss-300m-fortune.

6   "Mark Zuckerberg," *BrainyQuote,* accessed April 10, 2015, http://www.brainyquote.com/search_results.html?q=zuckerberg&pg=3.

7   Julia Fioretti, "Facebook 'tramples European privacy law': Belgian watchdog," *Reuters,* accessed May 18, 2015, http://www.reuters.com/article/2015/05/15/us-facebook-eu-privacy-idUSKBN0O0OXW20150515.

8   Julian Hattem, "Facebook claims 'a bug' made it track nonusers," *The Hill,* April 9, 2015, accessed April 10, 2015, http://thehill.com/policy/technology/238399-facebook-claims-a-bug-made-it-track-people-not-on-facebook.

9   "In the Matter of Facebook, Inc., a corporation," *Federal Trade Commission,* August 10, 2012, accessed March 9, 2015, https://www.ftc.gov/enforcement/cases-proceedings/092-3184/facebook-inc.

10 "Paul Lazarsfeld; Influence," *Wikipedia,* last updated May 4, 2015, accessed June 15, 2015http://en.wikipedia.org/wiki/Paul_Lazarsfeld#Influence.

11 Lawrence R. Samuel, *Freud on Madison Avenue* (Philadelphia: University of Pennsylvania Press, 2010), 21.

12 Jonathan R. Cole, "Paul F. Lazarsfeld: His Scholarly Journey," *Columbia University,* June 13, 2004, accessed February 20, 2015, http://www.columbia.edu/cu/univprof/jcole/_pdf/2004_Lazarsfeld.pdf.

    "Paul Lazarsfeld Guest Professorship," *University of Vienna,* accessed March 20, 2015, http://methods.univie.ac.at/paul-lazarsfeld-professorship/.

13 "Paul Lazarsfeld; Austria," *Wikipedia,* last updated May 4, 2015, accessed June 16, 2015, http://en.wikipedia.org/wiki/Paul_Lazarsfeld#Austria.

14 Samuel, *Freud on Madison Avenue,* 22.

15 "Paul Lazarsfeld," *Wikipedia,* last updated May 4, 2015, accessed June 15, 2015, http://en.wikipedia.org/wiki/Paul_Lazarsfeld.

16 "The Devil's Lure (?): Motivational Research 1934–1954," inn Robert A. Fullerton, *The Future of Marketing's Past,* accessed March 20, 2015, http://faculty.quinnipiac.edu/charm/CHARM%20proceedings/CHARM%20article%20archive%20pdf%20format/Volume%2012%202005/134%20fullerton.pdf, 135.

17 "Paul Lazarsfeld," *Wikipedia.*

"Mathematical sociology," *Wikipedia,* last updated May 26, 2015, accessed June 15, 2015.http://en.wikipedia.org/wiki/Mathematical_sociology.

18 James S. Coleman, *Introduction to Mathematical Sociology* (New York: Free Press of Glencoe/Macmillan, 1964), dedication page (unnumbered).

19 Ibid., 291.

20 "Social network analysis," *Wikipedia,* last updated 30 May 30, 2015, accessed June 15, 2015, http://en.wikipedia.org/wiki/Social_network_analysis.

21 Coleman, *Introduction to Mathematical Sociology,* 275, 299.

22 "Statement of Rights and Responsibilities," *Facebook,* accessed March 8, 2015, https://www.facebook.com/terms.php.

23. Steve Rosenbush, "Facebook Tests Software to Track Your Cursor on Screen," *Wall Street Journal,* October 30, 2013, last updated 7:15 a.m. ET, accessed May 22, 2014, http://blogs.wsj.com/cio/2013/10/30/facebook-considers-vast-increase-in-data-collection/.

24. Michal Kosinski, David Stillwell, and Thore Graepel, "Private traits and attributes are predictable from digital records of human behavior," *Proceedings of the National Academy of Science of the United States of America* 110 (approved February 12, 2013; received for review October 29, 2012), no. 15: 5,802–5, doi: 10.1073, http://www.pnas.org/content/110/15/5802.full.

25. Josh Gerstein and Stephanie Simon, "Who watches the watchers? Big Data goes unchecked," *Politico,* May 14, 2014, accessed March 31, 2015, http://www.politico.com/story/2014/05/big-data-beyond-the-nsa-106653.html.

26. "FTC Recommends Congress Require the Data Broker Industry to Be More Transparent and Give Consumers Greater Control Over Their Personal Information," *Federal Trade Commission,* May 27, 2014, http://www.ftc.gov/news-events/press-releases/2014/05/ftc-recommends-congress-require-data-broker-industry-be-more.

27. Ibid.

28. Ibid.

29. Phillip Bonacich and Philp Lu, *Introduction to Mathematical Sociology* (Princeton: Princeton University Press, 2012), 1.

30. "Social network," *Wikipedia,* accessed March 21, 2015, http://en.wikipedia.org/wiki/Social_network.

"Social Network Analysis," *Wikipedia,* last updated 30 May 30, 2015, accessed June 15, 2015, http://en.wikipedia.org/wiki/Social_network_analysis.

31 Sara Smyth, "Toll of social media on girls' mental health: Sexualised images fuelling rise in anxiety among pupils aged 11 to 13," *Daily Mail,* April 19, 2015, accessed April 19, 2015, http://www.dailymail.co.uk/news/article-3046222/Toll-social-media-girls-mental-health-Sexualised-images-fuelling-rise-anxiety-pupils-aged-11-13.html.

The Eyes Are the Windows to the Mind

1 Susana Martinez-Conde and Stephen L. Macknik, "Windows of the Mind," *Scientific American* (August 2007): 62, accessed March 31, 2015, http://smc.neuralcorrelate.com/files/publications/martinez-conde_macknik_sciam07.pdf.

Z. M. Hafed and J. J. Clark, "Microsaccades as an overt measure of covert attention shifts," *Vision Research* 42 (2002): 2,533–45, accessed March 31, 2015, http://www.cim.mcgill.ca/~clark/vmrl/web-content/papers/jjclark_vr_22_2002.pdf.

Ralf Engbert and Reinhold Kliegl, "Microsaccades uncover the orientation of covert attention," *Science Direct* 43, no. 9 (April 2003): 1,035–45, accessed March 31, 2015, http://www.sciencedirect.com/science/article/pii/S0042698903000841.

Ziad M. Hafed, "Alteration of Visual Perception prior to Microsaccades," *Neuron* 77 (2013): 775–86, accessed March 31, 2015, http://www.cnbc.cmu.edu/braingroup/papers/hafed_2013.pdf.

2 "Market Research," *SensoMotoric Instruments (SMI),* accessed March 4, 2015, http://www.smivision.com/en/

gaze-and-eye-tracking-systems/applications/market-research.html.

3   Kris Van Cleave, "Eye-tracking technology helps marketers and medical professionals alike," *WJLA,* accessed March 4, 2015, http://www.wjla.com/articles/2012/05/eye-tracking-technology-helps-marketers-and-medical-professionals-alike-75702.html.

4   "What the Eyes Reveal: 10 Messages My Pupils Are Sending You" *PsyBlog,* December 15, 2011, accessed April 9, 2015, http://www.spring.org.uk/2011/12/what-the-eyes-reveal-10-messages-my-pupils-are-sending-you.php.

5   Grant no. N00014-93-1-0525 from the Office of Naval Research, and grant no. F49620-97-1-0353 from the Air Force Office of Scientific Research. Sandra P. Marshall, US patent 6,090,051, "Statement of Government Rights: Method and apparatus for eye tracking and monitoring pupil dilation to evaluate cognitive activity," filed March 3, 1999, and issued July 18, 2000, *United States Patent and Trademark Office,* accessed February 25, 2015, http://patft.uspto.gov/netacgi/nph-Parser?Sect1=PTO1&Sect2=HITOFF&d=PALL&p=1&u=%2Fnetahtml%2FPTO%2Fsrchnum.htm&r=1&f=G&l=50&s1=6,090,051.PN.&OS=PN/6,090,051&RS=PN/6,090,051.

6   US patent 6,090,051, "Method and apparatus for eye tracking and monitoring pupil dilation to evaluate cognitive activity," *United States Patent and Trademark Office,* accessed March 31, 2015, http://patft.uspto.gov/netacgi/nph-Parser?Sect1=PTO1&Sect2=HITOFF&d=PALL&p=1&u=%2Fnetahtml%2FPTO%2Fsrchnum.htm&r=1&f=G&l=50&s1=6,090,051.PN.&OS=PN/6,090,051&RS=PN/6,090,051.

7   US patent 7,344,251, "Mental alertness level determination,"

United States Patent and Trademark Office, accessed April 1, 2015, http://patft.uspto.gov/netacgi/nph-Parser?Sect1=PTO2&Sect2=HITOFF&p=1&u=%2Fnetahtml%2FPTO%2Fsearch-bool.html&r=2&f=G&l=50&co1=AND&d=PTXT&s1=%22Index+Cognitive+Activity+%28ICA%29%22&OS=%22Index+Cognitive+Activity+%28ICA%29+%22&RS=%22Index+Cognitive+Activity+%28ICA%29%22.

8   Sandra P. Marshall, US patent 6,090,051, "Method and apparatus for eye tracking and monitoring pupil dilation to evaluate cognitive activity," filed March 3, 1999, and issued July 18, 2000, *United States Patent and Trademark Office,* accessed February 25, 2015, http://patft.uspto.gov/netacgi/nph-Parser?Sect1=PTO1&Sect2=HITOFF&d=PALL&p=1&u=%2Fnetahtml%2FPTO%2Fsrchnum.htm&r=1&f=G&l=50&s1=6,090,051.PN.&OS=PN/6,090,051&RS=PN/6,090,051.

9   "Software: Cognitive Workload," *EyeTracking, Inc.,* accessed February 25, 2015, http://www.eyetracking.com/Software/Cognitive-Workload.

10 US patent 7,438,418, "Mental alertness and mental proficiency level determination," *United States Patent and Trademark Office,* accessed April 1, 2015, http://patft.uspto.gov/netacgi/nph-Parser?Sect1=PTO1&Sect2=HITOFF&d=PALL&p=1&u=%2Fnetahtml%2FPTO%2Fsrchnum.htm&r=1&f=G&l=50&s1=7,438,418.PN.&OS=PN/7,438,418&RS=PN/7,438,418.

Marshall, US patent 6,090,051.

"Software: Cognitive Workload."

11 "The Index of Cognitive Activity: Measuring Cognitive Workload," Sandra P. Marshall, IEEE Human Factors Meeting, Scottsdale, Arizona, 2002, accessed May 3, 2013, http://

coursesite.uhcl.edu/hsh/peresSC/Classes/PSYC6419seminar/Index%20of%20cognitive%20activity.pdf (site discontinued). Now available for a fee at http://www.researchgate.net/publication/3973652 The Index of Cognitive Activity measuring cognitive workload.

12   US patent 7,438,418.

13   Ibid.

14   "Surprise! Eye Tracking Shows Men Look at Women and Men," *Eye Tracking Update,* August 21, 2010, accessed February 25, 2015, http://eyetrackingupdate.com/2010/08/21/surprise-eye-tracking-shows-men-women-men (site discontinued). A cached snapshot of the Web page as it appeared on March 22, 2015, is available at http://208.71.46.190/search/srpcache?p=Eye+Tracking+Shows+Men+Look+at+Women+and+Men&type=B111US837D20140925&fr=mcafee&ei=UTF-8&u=http://cc.bingj.com/cache.aspx?q=Eye+Tracking+Shows+Men+Look+at+Women+and+Men&d=4775425227756330&mkt=en-US&setlang=en-US&w=21yqYo1ID2x7nPioYlJf3EgQn4Lqn8rv&icp=1&.intl=us&sig=QiLOZkQFrf8M.9OKppCqQQ--.

15   "Eyetrak™ Visibility Research," *Visuality Group,* accessed February 25, 2015, http://www.visuality-group.co.uk/shopper-behaviour-and-research/eyetrak.

16   "Eye Tracking," *Tobii,* accessed March 4, 2014, http://www.mydelphi.eu/eye-tracking.html (site discontinued).

"Tobii Eye Tracking: See it from her point of view," *Acuity ETS,* accessed February 25, 2015, http://acuity-ets.com/downloads/tobii_mr_brochure_web.pdf.

17   "Eye Tracking: What catches your consumers' eyes?" *BDRC*

Continental, accessed March 4, 2014, http://www.bdrc-continental.com/technology-analytics/eye-tracking (site discontinued).

18 "Eye Tracking: A New Method for Marketers to Identify How You Really Feel about the Media You See," *Eye Tracking Update,* October 15, 2009, accessed February 25, 2015, http://eyetrackingupdate.com/2009/10/15/eye-tracking-a-new-method-for-marketers-to-identify-how-you-really-feel-about-the-media-you-see (site discontinued).

19 "Eye Tracking Demo at Shopper Marketing Live, May 19–20," *Objective Digital,* accessed May 8, 2013, http://www.objectivedigital.com/ot_may2011.htm (site discontinued).

20 Drew Harwell, "Pizza Hut wants to read your mind," *Washington Post,* December 1, 2014, accessed February 25, 2015, http://www.washingtonpost.com/blogs/the-switch/wp/2014/12/01/pizza-hut-wants-to-read-your-mind/.

21 "Eye tracking brings clarity to 'driving without awareness,'" *Tobii,* accessed February 25, 2015, http://www.tobii.com/en/eye-tracking-research/global/library/customer-cases/cognitive-psychology/attitudes/.

22 George Lakoff and Mark Johnson, *Philosophy in the Flesh: The Embodied Mind and Its Challenge to Western Thought* (New York: Basic Books, 1999), 13.

Gerald Zaltman, *How Consumers Think: Essential Insight into the Mind of the Market* (Boston: Harvard Business School Press, 2003), 58.

Adam L. Penenberg, "A. K. Pradeep, Mind Reader," *Fast Company,* August 10, 2011, accessed April 11, 2014, http://www.

fastcompany.com/1772167/ak-pradeep-mind-reader.

23 "Saccade," *Wikipedia,* last updated June 10, 2015, accessed June 15, 2015, http://en.wikipedia.org/wiki/Saccade.

24 Ann Marie Seward Barry, *Visual Intelligence, Perception, Image, and Manipulation in Visual Communication* (Albany: State University of New York Press, 1997), 32.

25 Susana Martinez-Conde and Stephen L. Macknik, "Mystery Solved," *Scientific American Mind* 22, no. 5, (November–December 2011): 54.

26 Susana Martinez-Conde, Jorge Otero-Millan, and Stephen L. Macknik, "The impact of microsaccades on vision: Toward a unified theory of saccadic function," *Nature Reviews Neuroscience* 14 (2013): 83–96, doi:10.1038/nrn3405, accessed April 9, 2014, http://www.nature.com/nrn/journal/v14/n2/full/nrn3405.html (subscription required).

27 Jorge Otero-Millan et al., "Saccades and microsaccades during visual fixation, exploration, and search: Foundations for a common saccadic generator," *Journal of Vision,* accessed April 11, 2015, http://www.journalofvision.org/content/8/14/21.long.

28 Martinez-Conde et al., "The impact of microsaccades on vision."

29 Marisa Carrasco, "Visual attention: The past 25 years," *Vision Research* 51, no. 13 (2011): 1,484–525, accessed April 11, 2015, http://www.sciencedirect.com/science/article/pii/S0042698911001544.

30 Ibid.

31 Martinez-Conde and Macknik, "Mystery Solved."

32 Susana Martinez-Conde, Stephen L. Macknik, Xoana G. Troncoso, and Thomas A. Dyarl, "Microsaccades Counteract Visual Fading during Fixation," *Neuron* 49 (January 19, 2006): 297–305, doi: 10.1016/j.neuron.2005.11.033, http://www.neuralcorrelate.com/smc/files/publications/martinez-conde_et_al_neuron06.pdf (subscription required).

Susana Martinez-Conde and Stephen L. Macknik, "Shifting Focus," *Scientific American Mind* 22, no. 5 (November–December 2011): 48–55, accessed April 21, 2015. http://www.scientificamerican.com/magazine/mind/2011/11-01/ (subscription required).

Martinez-Conde et al, "The impact of microsaccades on vision."

33 Susana Martinez-Conde, Stephen L. Macknik, Xoana G. Troncoso, and Thomas A. Dyarl, "Microsaccades Counteract Visual Fading during Fixation," *Neuron* 49 (January 19, 2006): 297–305, doi: 10.1016/j.neuron.2005.11.033, http://www.neuralcorrelate.com/smc/files/publications/martinez-conde_et_al_neuron06.pdf (subscription required).

34 Susana Martinez-Conde and Stephen L. Macknik, "Windows of the Mind," *Scientific American* (August 2007): 62, accessed March 31, 2015, http://smc.neuralcorrelate.com/files/publications/martinez-conde_macknik_sciam07.pdf.

35 Martinez-Conde et al, "The impact of microsaccades on vision."

36 *Oxford Dictionaries*, s.v. "covert," accessed May 27, 2013, http://oxforddictionaries.com/definition/english/covert.

37 Takemasa Yokoyama, Yasuki Noguchi, and Shinichi Kita, "Attentional shifts by gaze direction in voluntary orienting: Evidence from a microsaccade study," *Experimental Brain*

Research 223, no. 2 (September 23, 2012): 291–300, doi: 10.1007/s00221-012-3260-z, accessed April 11, 2015, http://www.ncbi.nlm.nih.gov/pmc/articles/PMC3475970.

38 Jochen Laubrock, Ralf Engbert, Martin Rolfs, and Reinhold Kliegl, "Microsaccades Are an Index of Covert Attention: Commentary on Horowitz, Fine, Fencsik, Yurgenson, and Wolfe (2007)," *Psychological Science,* accessed April 11, 2015, http://pss.sagepub.com/content/18/4/364.extract.

39 Martinez-Conde et al., "The impact of microsaccades on vision."

Ralf Engbert and Kliegl Reinhold, "Microsaccades uncover the orientation of covert attention," *Vision Research* 43, no. 9 (2003): 1,035–45, accessed April 11, 2015, doi:10.1016/S0042-6989(03)00084-1, http://www.sciencedirect.com/science/article/pii/S0042698903000841.

Ziad M. Hafed and James J. Clark, "Microsaccades as an overt measure of covert attention shifts," *Vision Research* 42 (2002): 2,533–45, accessed April 11, 2015, http://www.physiol-active-vision.uni-tuebingen.de/paper/hafed_vis_res_reprint2002.pdf.

40 Martinez-Conde et al., "The impact of microsaccades on vision."

41 Ibid.

42 Ibid.

Engbert and Reinhold, "Microsaccades uncover the orientation of covert attention."

Hafed and Clark, "Microsaccades as an overt measure of covert attention shifts."

43 Carrasco, "Visual attention."

44 Martinez-Conde and Macknik, "Shifting Focus."

45 Laubrock et al., "Microsaccades Are an Index of Covert Attention."

46 Martin Rolfs, Ralf Engber, and Reinhold Kliegl, "Crossmodal coupling of oculomotor control and spatial attention in vision and audition," *Experimental Brain Research* 166 (2005): 427–39, doi: 10.1007/s00221-005-2382-y, accessed April 21, 2015, http://openscience.uni-leipzig.de/index.php/mr2/article/view/76/63.

47 Engbert and Reinhold, "Microsaccades uncover the orientation of covert attention."

48 Martinez-Conde and Macknik, "Shifting Focus."

49 Ibid.

50 "Augmented Reality for Glass," *Augmented Reality for Glass*, accessed April 22, 2015, http://arforglass.org/.

51 "Google Glass makes doctors better surgeons: A Stanford study," *Surgery Academy,* September 29, 2014, accessed April 22, 2015, http://www.surgeryacade.my/2014/09/29/google-glass-makes-doctors-better-surgeons-a-stanford-study/.

52 Chris Wood, "Google eye tracking unlock patent revealed," *Gizmag,* August 8, 2012, accessed April 21, 2015, http://www.gizmag.com/google-eye-track-unlock-patent/23637.

53 Hayes Solos Raffle, Adrian Wong, and Ryan Geiss, US patent 8,235,529, "Unlocking a screen with eye tracking information," filed November 30, 2011, and issued August 7, 2012, *United States Patent and Trademark Office,* accessed April 21, 2015, http://patft.uspto.gov/netacgi/nph-Parser?Sect1=PTO1&Sect2=HITOFF&d=PALL&p=1&u=%2Fnetahtml%2FPTO%2Fsrchnum.htm&r=1&f=G&l=50&s1=8,235,529.PN.&OS=PN/8,235,5

29&RS=PN/8,235,529.

54 Hayes Solos Raffle, Adrian Wong, and Ryan Geiss, US patent 8,235,529, "Unlocking a screen with eye tracking information," filed November 30, 2011, and issued August 7, 2012, *United States Patent and Trademark Office,* accessed April 21, 2015, http://pdfpiw.uspto.gov/.piw?PageNum=0&docid=08235529&IDKey=DB61E66D7179%0D%0A&HomeUrl=http%3A%2F%2Fpatft.uspto.gov%2Fnetahtml%2FPTO%2Fpatimg.htm.

55 Ibid.

56 Martin Lindstrom, *Buyology* (New York: Crown Business, 2010), 20.

57 Emily Steel, "Companies scramble for consumer data," *Financial Times,* June 12, 2013, accessed March 3, 2015, http://www.ft.com/intl/cms/s/0/f0b6edc0-d342-11e2-b3ff-00144feab7de.html#axzz2W71Wg600 (subscription required).

58 Nick Pickels, "Google Glass: Orwellian surveillance with fluffier branding," *The Telegraph,* March 19, 2013, accessed April 21, 2015, http://www.telegraph.co.uk/technology/google/9939933/Google-Glass-Orwellian-surveillance-with-fluffier-branding.html.

59 Alice Truong, "This Google Glass App Will Detect Your Emotions, Then Relay Them Back to Retailers," *Fast Company,* March 6, 2014, accessed April 21, 2015, http://www.fastcompany.com/3027342/fast-feed/this-google-glass-app-will-detect-your-emotions-then-relay-them-back-to-retailers.

60 US patent 8,235,529.

61 "Calibration," *YouTube,* accessed April 21, 2015, http://www.youtube.com/watch?v=xXmbgb00Kxo.

62 Raffle et al., US patent 8,235,529.

63 Ibid.

64 Ibid.

65 That is, unless Google makes it clear in the user agreement, which is unlikely. In the case of Google's Gmail, the user agreement was not clear. It took legal action by Judge Lucy Koh to force Google to rewrite the user agreement regarding Gmail.

66 "Android (operating system)," *Wikipedia,* accessed April 21, 2015, http://en.wikipedia.org/wiki/Android_%28operating_system%29#Market_share.

67 "iPhone 6 Eye Tracking," *iPhone News,* accessed April 9, 2015, http://iphoneipadipod.com/apple/iphone-6-eye-tracking.html.

68 "Smartphone," *Wikipedia,* last updated June 10, 2015, accessed June 15, 2015, http://en.wikipedia.org/wiki/Smartphone#By_manufacturer.

69 Mark Prigg, "Now that's a Real iComputer: Samsung unveils eye and blink controlled machine to help the disabled get online," *Daily Mail,* November 26, 2014, accessed February 26, 2015, http://www.dailymail.co.uk/sciencetech/article-2850843/Samsung-unveils-EyeCan-machine-help-disabled-online.html#ixzz3KDpWzgpK.

70 "Samsung introduces EYECAN+, next-generation mouse for people with disabilities," *Samsung Village,* November 25, 2014, accessed April 21, 2015, http://www.samsungvillage.com/blog/2014/11/25/samsung-electronics-introduces-eyecan-next-generation-mouse-people-disabilities/.

71 David P. Julian, US patent 8,937,591, "Systems and methods

for counteracting a perceptual fading of a movable indicator," filed April 6, 2012, and issued January 20, 2015, *United States Patent and Trademark Office,* accessed April 21, 2015, http://patft.uspto.gov/netacgi/nph-Parser?Sect1=PTO2&Sect2=HITOFF&u=%2Fnetahtml%2FPTO%2Fsearch-adv.htm&r=2&p=1&f=G&l=50&d=PTXT&S1=%28345%2F157.CCLS.+AND+20150120.PD.%29&OS=ccl/345/157+and+isd/1/20/2015&RS=%28CCL/345/157+AND+ISD/20150120%29, claims 10 and 27.

72 "General Information Concerning Patents, Specification [Description and Claims]," *US Patent and Trademark Office,* November 2011, accessed April 21, 2015, http://www.uspto.gov/patents/resources/general_info_concerning_patents.jsp#heading-17.

73 Raffle et al., "Unlocking a screen with eye tracking information."

74 Mark Prigg, "Glass without the glasses: Google patents smart contact lens system with a camera built in," *Daily Mail,* April 14, 2014, accessed 21 April 21, 2015, http://www.dailymail.co.uk/sciencetech/article-2604543/Glass-without-glasses-Google-patents-smart-contact-lens-CAMERA-built-in.html.

Electronic Mind Readers

1 Mary Carmichael, "Neuromarketing: Is it coming to a lab near you?" *PBS,* accessed April 21, 2015, http://www.pbs.org/wgbh/pages/frontline/shows/persuaders/etc/neuro.html.

2 George Lakoff and Mark Johnson, *Philosophy in the Flesh: The Embodied Mind and Its Challenge to Western Thought* (New York: Basic Books, 1999), 13.

Gerald Zaltman, *How Consumers Think: Essential Insight into the Mind of the Market* (Boston: Harvard Business School Press, 2003), 58.

Adam L. Penenberg, "A. K. Pradeep, Mind Reader," *Fast Company,* August 10, 2011, accessed April 21, 2015, http://www.fastcompany.com/1772167/ak-pradeep-mind-reader.

3   "Neuromarketing," *Wikipedia,* last updated April 13, 2015, accessed June 15, 2015, http://en.wikipedia.org/wiki/Neuromarketing.

4   "Electroencephalography," *Wikipedia,* last updated June 13, 2015., accessed June 15, 2015, http://en.wikipedia.org/wiki/Electroencephalography.

5   *Emotiv,* accessed January 31, 2015, http://www.emotiv.com/.

6   Ibid.

7   *Puzzlebox.io,* accessed January 31, 2015, http://puzzlebox.io/orbit/.

8   Erica Warp, "NeuroDisco," *Erica Warp,* accessed January 31, 2015, http://ericawarp.com/?projects=neurodisco.

9   "Steady state topography," *Wikipedia,* last updated May 26, 2015, accessed June 15, 2015, http://en.wikipedia.org/wiki/Steady_state_topography.

10 Martin Lindstrom, *Buyology* (New York: Crown Business, 2010), 24.

11 "Introduction to Buyology: The Truth and Lies about Why We Buy," *YouTube,* accessed January 30, 2015, https://www.youtube.com/watch?v=VVCVQ7TYC04&x-yt-ts=1422579428&x-yt-cl=85114404#t=230 (three minutes and forty-five seconds into

the video).

12. *Neuro-Insight,* accessed January 31, 2015, http://www.neuro-insight.com/.

13. Lindstrom, *Buyology,* 34.

Can this "Mechanism" Accurately Reflect Your Thoughts?

1. Carl Zimmer, "Secrets of the Brain," *National Geographic* 225, no. 2 (February 2014): 55.

2. Anantha Pradeep, Robert T. Knight, and Ramachandran Gurumoorthy, US patent 8,494,905, filed June 6, 2008, and issued July 23, 2013. Column 3, lines 47–59, "accurately reflect a subject's actual thoughts. … Consequently, the techniques and mechanisms of the present invention intelligently blend multiple modes such as EEG and fMRI to more accurately assess effectiveness of stimulus materials."

3. Anantha Pradeep, US patent 8,494,905, filed June 6, 2008, and issued July 23, 2013, *United States Patent and Trademark Office,* accessed May 20, 2014, http://pdfpiw.uspto.gov/.piw?PageNum=0&docid=08494905&IDKey=AFD0F456F6C5%0D%0A&HomeUrl=http%3A%2F%2Fpatft.uspto.gov%2Fnetahtml%2FPTO%2Fpatimg.htm.

4. *Nielsen,* accessed January 27, 2015, http://www.nielsen.com/us/en/solutions/measurement.html.

5. These word counts include words in references cited within this patent document. It is very rare for patents to extensively reference scientific and business articles separate from the "prior

art" of previous patents. Furthermore, this patent is unique in many other ways. It took over six years from the filing of the provisional application to the issuance date, which is much longer than average. It also includes a very long list of office actions by the USPTO as well as letters from several foreign patent offices. The paper trail suggests that the legal efforts to obtain this patent were very costly and very long. These many unusual factors suggest how profitable this patent was deemed to be by the assignee—Nielsen Company.

6   Margaret Talbot, "Duped," *New Yorker*, July 2, 2007, accessed April 22, 2015, http://www.newyorker.com/magazine/2007/07/02/duped.

7   "The Pleasure Centres Affected by Drugs, *The Brain, from Top to Bottom*, accessed March 6, 2015, http://thebrain.mcgill.ca/flash/a/a_03/a_03_cr/a_03_cr_par/a_03_cr_par.html.

"Nucleus accumbens," *Wikipedia*, last updated April 17, 2015, accessed June 15, 2015, http://en.wikipedia.org/wiki/Nucleus_accumbens.

8   Martin Lindstrom, *Buyology* (New York, Crown Business, 2010), 14.

9   Brian Knutson, "Visualizing Desire," *YouTube*, uploaded January 22, 2009, accessed March 17, 2015, https://www.youtube.com/watch?v=CUK8D-kXofE.

10 Martin Lindstrom, *Buyology* (New York: Crown Business, 2010), 15.

11 *Facial Action Coding System*, accessed May 24, 2014, http://facialactioncoding.com/.

12 David Talbot, "Startup Gets Computers to Read Faces, Seeks

Purpose beyond Ads," *MIT Technology Review*, accessed March 4, 2015, http://www.technologyreview.com/news/519656/startup-gets-computers-to-read-faces-seeks-purpose-beyond-ads/.

13 Anantha Pradeep, US patent 8,494,905, filed June 6, 2008, and issued July 23, 2013, *United States Patent and Trademark Office*, accessed May 20, 2014, http://pdfpiw.uspto.gov/.piw?PageNum=0&docid=08494905&IDKey=AFD0F456F6C5%0D%0A&HomeUrl=http%3A%2F%2Fpatft.uspto.gov%2Fnetahtml%2FPTO%2Fpatimg.htm, column 7, lines 30–35.

14 Ibid., column 5, lines 29–38; column 8, lines 1–10; and column 12, lines 12–18.

15 Pradeep et al., US patent 8,494,905, column 1, lines 34–46 and column 3, lines 50–55.

16 Pradeep et al., US patent 8,494,905, column 3, lines 33–55:

> Some conventional mechanisms cite a particular neurological or neurophysiological measurement characteristic as indicating a particular thought, feeling, mental state, or ability. For example, one mechanism purports that the contraction of a particular facial muscle indicates the presence of a particular emotion. Others measure general activity in particular areas of the brain and suggest that activity in one portion may suggest lying while activity in another portion may suggest truthfulness. However, these mechanisms are severely limited in their ability to accurately reflect a subject's actual thoughts. It is recognized that a particular region of the brain cannot be mapped to a particular thought. Similarly, a particular eye movement cannot be

mapped to a particular emotion. Even when there is a strong correlation between a particular measured characteristic and a thought, feeling, or mental state, the correlations are not perfect, leading to a large number of false positives and false negatives. ... Consequently, the techniques and mechanisms of the present invention intelligently blend multiple modes such as EEG and fMRI to more accurately assess effectiveness of stimulus materials.

17 "Alpha Brain Waves Background Information," *Brain Waves Blog,* accessed March 6, 2015, http://www.brainwavesblog.com/tag/alpha-waves/.

18 "Gamma Brain Waves Information," *Brain Waves Blog,* accessed March 6, 2015, http://www.brainwavesblog.com/gamma-brain-waves-information/.

19 "Alpha Brain Waves Background Information."

20 "What Are Gamma Brain Waves?" *Brain Waves Blog,* accessed March 6, 2015, http://www.brainwavesblog.com/gamma-brain-waves-information/.

21 Shelley Carson, "The Creative Mind," *Scientific American Mind Special Report* (2014): 31.

22 Hannah Devlin, "What Is Functional Magnetic Resonance Imaging (fMRI)?" *Psych Central,* accessed March 7, 2015, http://psychcentral.com/lib/what-is-functional-magnetic-resonance-imaging-fmri/0001056.

23 Pradeep et al., US patent 8,494,905, column 4, lines 45–47: "A variety of modalities can be used including EEG, GSR, EKG, pupillary dilation, EOG, eye tracking, facial emotion encoding,

reaction time, etc."

24. Nick Saint, "Google CEO: 'We Know Where You Are. We Know Where You've Been. We Can More or Less Know What You're Thinking About,'" *Business Insider,* October 4, 2010, accessed March 25, 2015, http://www.businessinsider.com/eric-schmidt-we-know-where-you-are-we-know-where-youve-been-we-can-more-or-less-know-what-youre-thinking-about-2010-10.

25. Derek Thompson, "Google's CEO: 'The Laws Are Written by Lobbyists,'" *The Atlantic,* October 1, 2010, accessed April 21, 2015, http://www.theatlantic.com/technology/archive/2010/10/googles-ceo-the-laws-are-written-by-lobbyists/63908/.

26. If you are not reading this on your smartphone, consider if your smartphone was monitoring your eye movements, facial expressions and voice as you read blogs, news, email, tweets, etc.

27. Pradeep et al., US patent 8,494,905, column 5, lines 19–22.

28. "Data Brokers: A Call for Transparency and Accountability," *Federal Trade Commission,* May 27, 2014, accessed April 21, 2015, http://www.ftc.gov/system/files/documents/reports/data-brokers-call-transparency-accountability-report-federal-trade-commission-may-2014/140527databrokerreport.pdf, 27.

29. Ibid., 28.

30. "HTTP cookie," *Wikipedia,* last updated June 15, 2015, accessed June 15, 2015, http://en.wikipedia.org/wiki/HTTP_cookie.

31. "Data Brokers," 29.

32. "Data Brokers," 31.

33. Ibid.

"Wearables": For Your Health or for Their Profit?

1. Alistair Barr, "New Google Glass Version Coming this Year," *Wall Street Journal,* January 15, 2015, accessed March 2, 2015, http://www.wsj.com/video/new-google-glass-version-coming-this-year/B69DDCD7-1C14-4556-B5D9-26487E3CF49D.html (three minutes and thirty-five seconds into the video).

2. "The wear, why and how," *The Economist,* March 14, 2015, accessed April 12, 2015, http://www.economist.com/news/business/21646225-smartwatches-and-other-wearable-devices-become-mainstream-products-will-take-more?zid=291&ah=906e69ad01d2ee51960100b7fa502595.

3. Bill Rigby, "Exclusive: Six percent of U.S. adults plan to buy Apple Watch—Reuters/Ipsos poll," *Reuters,* April 15, 2015, accessed April 15, 2015, http://www.reuters.com/article/2015/04/15/us-apple-watch-idUSKBN0N628820150415.

4. "Google's Top Tech Search Trends of 2014—Digits," *Wall Street Journal,* accessed March 6, 2015, http://www.wsj.com/video/google-top-tech-search-trends-of-2014-digits/30ED1740-74EF-43FC-A549-E0D8B2024407.html?KEYWORDS=fear+psychology. (Health trackers and smart watches are shown in the video between time markers 3:10 and 6:07.)

5. Jon Phillips, "The 10 most likely sensors in a 10-sensor Apple smartwatch," *TechHive,* June 21, 2014, accessed March 2, 2015, http://www.techhive.com/article/2366126/the-10-most-likely-sensors-in-a-10-sensor-apple-smartwatch.html.

6. "Apple Watch," *Wikipedia,* last updated June 15, 2015, accessed June 15, 2015, http://en.wikipedia.org/wiki/Apple_Watch#Technology.

7. *Apple,* accessed March 2, 2015, http://www.apple.com/watch/technology/#familiar.

8. Septime Meunier, "Health checks by smartphone raise privacy fears," *Yahoo! News,* March 5, 2015, accessed March 5, 2015, http://news.yahoo.com/health-checks-smartphone-raise-privacy-fears-080635403.html;_ylt=AwrSyCOdk_hUcAsATy7QtDMD.

9. Chriss W. Street, "i-Watch Could Let Managers Snoop on Employees," *Breitbart,* April 6, 2015, accessed April 6, 2015, http://www.breitbart.com/california/2015/04/06/i-watch-could-let-managers-snoop-on-employees/.

Someday Soon, Maybe Very Soon

1. Bruce Schneier, *Data and Goliath: The Hidden Battle to Collect Your Data and Control You* (New York, W. W. Norton, 2015), 6.

2. This is the name for a notional company. A quick Internet search revealed that there were not any real companies associated with this name. Any resemblance to a real company is coincidental.

3. Carolyn Giardina, "'Furious 7' and How Peter Jackson's Weta Created Digital Paul Walker," *The Hollywood Reporter,* March 25, 2015, accessed March 25, 2015, http://www.hollywoodreporter.com/behind-screen/furious-7-how-peter-jacksons-784157.

4. *Autodesk 123D,* accessed March 19, 2015, http://www.123dapp.com/catch.

   Lydia Sloan Cline, *3D Printing with Autodesk 123D, Tinkercad, and Makerbot* (New York: McGraw Hill, 2015), 87–133.

   Jianguo Li, Eric Li, Yurong Chen, and Lin Xu, "Visual 3D Modeling from Images and Videos," *Intel Labs* (China), June 3,

2010, accessed February 18, 2015, https://16cbeb23-a-62cb3a1a-s-sites.googlegroups.com/site/leeplus/3DTR.pdf?attachauth=ANoY7cq2WWmGGpnGHFNxYdWzyEeNb4aOj5xhkQUsg43k8OgtU4H7f97QrXnkmXiJ5HODn9yok-zrO-poJsp3ayh4OP7zoP6xLWP4GEUswVosOowG7f2abS6a4cyo653qjcD8rVieTDcIbefcofrf-LqzAhwhbHSjbZ_bwl9-BATeE5oDUcvCB3JPPCJDVQueImLKY5p4KIIU&attredirects=0 and https://docs.google.com/file/d/0B8_ZlPz5Dh7mUHlCM0VCLXhUYnEzV1lGUDFhTkxVUQ/edit?pli=1.

5   This is the name for a notional company. A quick Internet search revealed that there are not any real companies associated with this name. Any resemblance to a real company is coincidental.

6   *The Sash Company,* accessed February 6, 2015, https://www.thesashcompany.com/missamerica.php.

7   Good for you for catching the intentional wordplay. In the earlier instances in this book, *crack* is used as an adjective meaning "first-rate." In this situation, *crack* is used to indicate a form of cocaine. Some marketing psychologists strive to find ways to stimulate the nucleus accumbens just as cocaine stimulates the nucleus accumbens. Martin Lindstrom, *Buyology* (New York: Crown Business, 2010), 14–15. This comparison may elicit howls of protest from marketing psychologists, but you can judge for yourself.

8   Kyle Smith, "Google controls what we buy, the news we read—and Obama's policies," *New York Post,* March 28, 2015, accessed March 30, 2015, http://nypost.com/2015/03/28/google-controls-what-we-buy-the-news-we-read-and-obamas-policies/.

Legal and Ethical Issues

1. Nick Saint, "Google CEO: 'We Know Where You Are. We Know Where You've Been. We Can More or Less Know What You're Thinking About,'" *Business Insider,* October 4, 2010, accessed March 25, 2015, http://www.businessinsider.com/eric-schmidt-we-know-where-you-are-we-know-where-youve-been-we-can-more-or-less-know-what-youre-thinking-about-2010-10.

2. Elaine Ganley, Angela Charlton, and Frank Jordans, "Germanwings Co-Pilot Researched Suicide Methods, Cockpit Security: Prosecutors," *Huffington Post,* posted April 2, 2015, 10:15 a.m. EDT, last modified April 2, 2015, 4:59 p.m. EDT, accessed April 3, 2015, http://www.huffingtonpost.com/2015/04/02/andreas-lubitz-suicide-methods_n_6992420.html.

   Tom Käckenhoff and Jean-Francois Rosnoblet, "REFILE-UPDATE 3—German pilot researched suicide, cockpit doors; second black box found," *Reuters,* April 2, 2015, 2:16 p.m. EDT, accessed April 4, 2015, http://www.reuters.com/article/2015/04/02/france-crash-recorder-update-3-tv-pixcor-idUSL6N0WZ31T20150402.

1. Mary Madden, "Public Perceptions of Privacy and Security in the Post-Snowden Era," *Pew Research Center,* November 12, 2014, accessed January 28, 2015, http://pewrsr.ch/1wlEbYR.

   The following is a direction quotation from the article:

   Yet, even as Americans express concern about government access to their data, they feel as though government could do more to regulate what advertisers do with their personal information:

- 80% of adults "agree" or "strongly agree" that Americans should be concerned about the government's monitoring of phone calls and internet communications. Just 18% "disagree" or "strongly disagree" with that notion.

- 64% believe the government should do more to regulate advertisers, compared with 34% who think the government should not get more involved.

- Only 36% "agree" or "strongly agree" with the statement: "It is a good thing for society if people believe that someone is keeping an eye on the things that they do online."

2   Jon Buys, "DuckDuckGo: A New Search Engine Built from Open Source," *OStatic,* July 10, 2010, accessed April 1, 2015, http://ostatic.com/blog/duckduckgo-a-new-search-engine-built-from-open-source.

Bruce Schneier, *Data and Goliath: The Hidden Battle to Collect Your Data and Control You* (New York: W. W. Norton, 2015), 124.

3   *Ello,* accessed April 5, 2015, https://ello.co/wtf/post/about.

Schneier, *Data and Goliath,* 124.

4   "Avery® Assorted Removable Color Coding Labels 5795, 1/4" Round, Pack of 760," accessed May 29, 2015, http://www.avery.com/avery/en_us/Products/Labels/Identification-Labels/Color-Coding-Labels_05795.htm?N=0&Ns=&refchannel=c042fd03ab30a110VgnVCM1000002118140aRCRD.

5   "Avery® Dark Blue Removable Print or Write Color Coding Labels for Laser and Inkjet Printers 5469, 3/4" Round, Pack of 1008," accessed May 29, 2015, http://www.avery.

com/avery/en_us/Products/Labels/Identification-Labels/Print-or-Write-Round-Color-Coding-Labels_05469.htm?N=0&Ns=&refchannel=c042fd03ab30a110VgnVCM100000 2118140aRCRD.

Glossary

[1] *Merriam-Webster,* s.v. "algorithm," accessed March 26, 2015, http://www.merriam-webster.com/dictionary/algorithm.

[2] Keith J. Kaplan, MD, "ZEISS Axio Scan.Z1 Advanced Digital Imaging System Improves Pathology Research," *Digital Pathology Blog,* March 7, 2014, accessed March 26, 2015, http://tissuepathology.com/2014/03/07/zeiss-axio-scan-z1-advanced-digital-imaging-system-improves-pathology-research/#axzz3bXKlm100 /.

[3] "Using Artificial Intelligence to Write Self-Modifying/Improving Programs," *Primary Objects,* accessed March 26, 2015, http://www.primaryobjects.com/CMS/Article149.

"Artificial Intelligence Wikipedia-based Free Textbook," *SourceForge,* accessed March 26, 2015, http://mind.sourceforge.net/aisteps.html.

[4] News staff, "Sabina, a Robot Domestic Learns When You Show Her," *Science 2.0,* April 2, 2015, accessed April 3, 2015, http://www.science20.com/news_articles/sabina_a_robot_domestic_learns_when_you_show_her-154512.

[5] "Anonymous-entity resolution," *ITLaw* wiki, accessed April 8, 2015, http://itlaw.wikia.com/wiki/Anonymous-entity_resolution.

[6] John McCarthy, "What Is Artificial Intelligence?" *Stanford University,* revised November 12, 2007, accessed March 26, 2015,

http://www-formal.stanford.edu/jmc/whatisai/node1.html.

7   Alistair Barr, "Google buys artificial intelligence start-up DeepMind," *USA Today*, January 27, 2014, accessed March 26, 2015, http://www.usatoday.com/story/tech/2014/01/27/google-deepmind-artificial-intelligence/4943049/.

8   *Merriam-Webster*, s.v. "autonomous," accessed March 26, 2015, http://www.merriam-webster.com/dictionary/autonomous.

9   Susana Martinez-Conde and Stephen L. Macknik, "Windows of the Mind," *Scientific American* (August 2007): 62, accessed March 31, 2015, http://smc.neuralcorrelate.com/files/publications/martinez-conde_macknik_sciam07.pdf.

Z. M. Hafed and J. J. Clark, "Microsaccades as an overt measure of covert attention shifts," *Vision Research* 42 (2002): 2,533–45, accessed March 31, 2015, http://www.cim.mcgill.ca/~clark/vmrl/web-content/papers/jjclark_vr_22_2002.pdf.

Ralf Engbert and Reinhold Kliegl, "Microsaccades uncover the orientation of covert attention," *Science Direct* 43, no. 9 (April 2003): 1,035–45, accessed March 31, 2015, http://www.sciencedirect.com/science/article/pii/S0042698903000841.

Ziad M. Hafed, "Alteration of Visual Perception prior to Microsaccades," *Neuron* 77 (2013): 775–86, accessed March 31, 2015, http://www.cnbc.cmu.edu/braingroup/papers/hafed_2013.pdf.

10 Simon Singh, *The Code Book* (New York: Anchor Books, 1999), 1–44.

11 "Dick Tracy," *Wikipedia*, accessed April 15, 2015, http://en.wikipedia.org/wiki/Dick_Tracy.

12 Takatoshi Hikida, Satoshi Yawata, Takashi Yamaguchi, Teruko

Danjo, Toshikuni Sasaoka, Yanyan Wang, and Shigetada Nakanishi, "Pathway-specific modulation of nucleus accumbens in reward and aversive behavior via selective transmitter receptors," *Proceedings of the National Academy of Sciences of the United States of America* 110, no. 1 (January 2, 2013): 342–7. Published online on December 17, 2012, accessed April 6, 2015, doi: 10.1073/pnas.1220358110, http://www.ncbi.nlm.nih.gov/pmc/articles/PMC3538201.

Dirk Hanson, "Addiction Inbox," February 10, 2010, adapted from Dirk Hanson, The Chemical Carousel: *What Science Tells Us about* Beating Addiction (Charleston, SC: BookSurge Publishing, 2009), accessed April 6, 2015, http://addiction-dirkh.blogspot.com/2010/02/nucleus-accumbens.html.

[13] *Merriam-Webster,* s.v. "nucleus accumbens," accessed March 26, 2015, http://www.merriam-webster.com/medical/nucleus%20accumbens.

[14] *Dictionary.com,* s.v. "neuromarketing," accessed April 6, 2015, http://dictionary.reference.com/browse/neuromarketing.

[15] "Neuromarketing," *Wikipedia,* last updated April 13, 2015, accessed June 15, 2015, http://en.wikipedia.org/wiki/Neuromarketing.

[16] *The Free Medical Dictionary,* s.v. "psychophysiological," accessed April 7, 2015, http://medical-dictionary.thefreedictionary.com/Psychophysiological.

[17] *The Free Medical Dictionary,* s.v. "saccades," accessed March 26, 2015, http://medical-dictionary.thefreedictionary.com/Saccades.

[18] "Steady state topography," *Wikipedia,* last updated May 26, 2015, accessed June 15, 2015, http://en.wikipedia.org/wiki/Steady_state_topography.

www.ingramcontent.com/pod-product-compliance
Lightning Source LLC
Chambersburg PA
CBHW020737180526
45163CB00001B/263